KT-441-395

Contents

Guidance Note 4

Protection Against Fire

17th IEE Wiring Regulations Seventeenth Edition
BS 7671:2008 Requirements for Electrical Installations

Published by The Institution of Engineering and Technology, London, United Kingdom

The Institution of Engineering and Technology is registered as a Charity in England & Wales (no. 211014) and Scotland (no. SCO38698).

The Institution of Engineering and Technology is the new institution formed by the joining together of the IEE (The Institution of Electrical Engineers) and the IIE (The Institution of Incorporated Engineers). The new Institution is the inheritor of the IEE brand and all its products and services, such as this one, which we hope you will find useful. The IEE is a registered trademark of the Institution of Engineering and Technology.

First published 1992 (0 85296 539 7)
Reprinted (with amendments) 1993
Reprinted (with minor amendments) 1994
Second edition (incorporating Amendment No. 1 to BS 7671:1992) 1995 (0 85296 868 X)
Third edition (incorporating Amendment No. 2 to BS 7671:1992) 1998 (0 85296 957 0)
Fourth edition (incorporating Amendment No. 1 to BS 7671:2001) 2003 (0 85296 992 9)
Reprinted (incorporating Amendment No. 2 to BS 7671:2001) 2004
Fifth edition (incorporating BS 7671:2008) 2009 (978-0-86341-858-7)

Copies of this publication may be obtained from:
The Institution of Engineering and Technology
PO Box 96
Stevenage
SG1 2SD, UK
Tel: +44 (0)1438 767328
Email: sales@theiet.org
www.theiet.org/publishing/books/wir-reg/

While the author, publisher and contributors believe that the information and guidance given in this work are correct, all parties must rely upon their own skill and judgement when making use of them. The author, publisher and contributors do not assume any liability to anyone for any loss or damage caused by any error or omission in the work, whether such an error or omission is the result of negligence or any other cause. Where reference is made to legislation it is not to be considered as legal advice. Any and all such liability is disclaimed.

ISBN 978-0-86341-858-7

Typeset in the UK by The Institution of Engineering and Technology
Printed in the UK by Printwright Ltd, Ipswich

Cooperating organisations

The Institution of Engineering and Technology acknowledges the contribution made by the following organisations in the preparation of this Guidance Note.

British Cables Association
J.M.R. Hagger BTech(Hons) AMIMMM
C.K. Reed IEng MIET

British Electrotechnical & Allied Manufacturers Association Ltd
P.D. Galbraith IEng MIET

BEAMA Installation Ltd
Eur Ing M.H. Mullins BA CEng FIEE FIIE

Electrical Contractors' Association
H.R. Lovegrove IEng FIET
Eur Ing L. Markwell MSc BSc CEng MIET MCIBSE LCGI

Electrical Safety Council
H. Goodenough BSc CEng FIEE FCIBSE FSLL

Health and Safety Executive
N. Gove MEng CEng MIEE
K. Morton BSc CEng MIEE

Institution of Engineering and Technology
P.R.L. Cook CEng FIEE MCIBSE
G.D. Cronshaw IEng FIET
P.E. Donnachie BSc CEng FIET
J.F. Elliott BSc(Hons) PG Cert IEng MIEE (Editor)
J. Simmons CEng FIEE

Lighting Association
L. Barling

SELECT (Electrical Contractors' Association of Scotland)
R.M. Cairney IEng MIET
D. Millar IEng MIET MILE

Acknowledgements

References to British Standards, CENELEC Harmonization Documents and International Electrotechnical Commission standards are made with the kind permission of BSI. Complete copies can be obtained by post from:

BSI Customer Services
389 Chiswick High Road
London W4 4AL
Tel: +44 (0)20 8996 9001
Fax: +44 (0)20 8996 7001

BSI also maintains stocks of international and foreign standards, with many English translations. Up-to-date information on BSI standards can be obtained from the BSI website: www.bsi-global.com

Complete copies of Approved Document B and the other Approved Documents are downloadable from the planning portal website: www.planningportal.gov.uk

The Building Standards Division of the Scottish Government, Directorate for the Built Environment, website can be found at: www.sbsa.gov.uk

The Northern Ireland Building Regulations website can be found at:
www.buildingregulationsni.gov.uk

The illustrations within this publication were provided by Rod Farquhar Design:
www.rodfarquhar.co.uk

Cover design and illustration were created by The Page Design:
www.thepagedesign.co.uk

The flameguard downlighter and insulation support box images were supplied courtesy of CLICK Scolmore: www.scolmore.com

Copies of Health and Safety Executive documents and approved codes of practice (ACOP) can be obtained from:

HSE Books
PO Box 1999
Sudbury, Suffolk CO10 2WA
Tel: +44 (0)1787 881165
Email: hsebooks@prolog.uk.com
Web: www.hsebooks.com

The HSE website is www.hse.gov.uk

Preface

This Guidance Note is part of a series issued by the Institution of Engineering and Technology to explain and enlarge upon the requirements in BS 7671:2008, the 17th Edition of the IEE Wiring Regulations.

Note that this Guidance Note does not ensure compliance with BS 7671. It is intended to explain some of the requirements of BS 7671, but readers should always consult BS 7671 to satisfy themselves of compliance.

The scope generally follows that of BS 7671; the relevant Regulations and Appendices are noted in the margin. Some Guidance Notes also contain material not included in BS 7671:2008 but which was included in earlier editions of the Wiring Regulations. All of the Guidance Notes contain references to other relevant sources of information.

Electrical installations in the United Kingdom that comply with BS 7671 are likely to satisfy the relevant aspects of Statutory Regulations such as the Electricity at Work Regulations 1989, but this cannot be guaranteed. It is stressed that it is essential to establish which Statutory and other Regulations apply and to install accordingly. For example, an installation in premises subject to licensing may have requirements different from, or additional to, BS 7671 and these will take precedence.

Introduction

Protection against fire resulting from the installation and use of electrical installations has been necessary ever since the use of electricity was first introduced into buildings. In fact, it is worth noting that the first edition of the *Wiring Regulations*, introduced in 1882, was called the *Rules and Regulations for the Prevention of Fire Risks Arising from Electric Lighting.*

This Guidance Note is concerned primarily with Chapter 42 and those other parts of BS 7671 which concern protection against thermal effects. It does not attempt to deal comprehensively with the safety of personnel in the event of fire, since this is beyond the scope of BS 7671.

Neither BS 7671 nor the Guidance Notes are design guides. It is essential to prepare a full design and specification prior to commencement or alteration of an electrical installation. Compliance with the relevant standards should be required.

The design and specification should set out the requirements and provide sufficient information to enable competent persons to carry out the installation and to commission it. The specification must include a description of how the system is to operate and all the design and operational parameters. It must provide for all the commissioning procedures that will be required and for the provision of adequate information to the user. This will be by means of an operational and maintenance manual or schedule, 514.9 and 'as fitted' drawings if necessary.

It must be noted that it is a matter of contract as to which person or organisation is responsible for the production of the parts of the design, specification, construction and verification of the installation and any operational information.

The persons or organisations who may be concerned in the preparation of the works include:

> The Designer
> The CDM Coordinator
> The Installer
> The Supplier of Electricity (Distributor)
> The Installation Owner (Client) and/or User
> The Architect
> The Fire Prevention Officer
> All Regulatory Authorities
> Any Licensing Authority
> The Health and Safety Executive

In producing the design, advice should be sought from the installation owner and/or 132.1 user as to the intended use. Often, as in a speculative building, the intended use is unknown. The specification and/or the operational manual must set out the basis of use for which the installation is suitable.

133.1
Section 511 Precise details of each item of equipment should be obtained from the manufacturer and/or supplier and compliance with appropriate standards confirmed.

The operational manual must include a description of how the system as installed is to operate and all commissioning records. The manual should also include manufacturers' technical data for all items of switchgear, luminaires, accessories etc. and any special instructions that may be needed.

The Health and Safety at Work etc. Act 1974 Section 6 and the Construction (Design and Management) Regulations 2007 are concerned with the provision of information, and guidance on the preparation of technical manuals is given in the BS 4884 series *Technical manuals* and the BS 4940 series *Technical information on constructional products and services*. The size and complexity of the installation will dictate the nature and extent of the manual.

Statutory requirements

1.1 General

A number of enactments and statutory instruments including the Electricity at Work Regulations 1989 (made under the Health and Safety at Work etc. Act 1974) deal with fire risks, prevention of fire and fire precautions.

This Guidance Note is not intended to provide an exhaustive treatment of the legislation concerned with fire, but deals only with those situations referred to in BS 7671. Thus, certain specialised installations listed in Regulation 110.2 are excluded.

110.2

1.2 Statutory Regulations

Appx 2

1.2.1 The Electricity at Work Regulations 1989 as amended

The Electricity at Work Regulations are general in their application and refer throughout to 'danger' and 'injury'. Danger is defined as risk of 'injury' and injury is defined in terms of certain classes of potential harm to persons. Injury is stated to mean death or injury to persons from:

▶ electric shock
▶ electric burn
▶ electrical explosion or arcing
▶ fire or explosion initiated by electrical energy.

The Health and Safety Executive *Memorandum of Guidance on the Electricity at Work Regulations 1989* (HSR25) is essential reading for all concerned with electrical installations.

HSR25

1.2.2 The Electricity Safety, Quality and Continuity Regulations 2002 as amended

The definition of 'danger' in these Regulations includes 'danger to health or danger to life or limb from electric shock, burn, injury or mechanical movement to persons, livestock or domestic animals, or from fire or explosion, attendant upon the generation, transmission, transformation, distribution or use of energy'.

1.2.3 The Management of Health and Safety at Work Regulations 1999

These Regulations place general duties on employers to assess risks to the health and safety of employees and others and take managerial action to minimise these risks, including:

▶ implementing preventive measures
▶ providing health surveillance
▶ appointing competent people

- ▶ setting up procedures
- ▶ providing information
- ▶ training etc.

Note: These Regulations have been subject to some revocations under the Regulatory Reform (Fire Safety) Order 2005. See Schedule 5 of the Order for full details.

1.2.4 The Provision and Use of Work Equipment Regulations 1998

These Regulations require employers to ensure that work equipment is suitable for the purpose.

The Regulations also require that work equipment is maintained in an efficient state, in efficient working order and in good repair.

1.2.5 The Construction (Design and Management) Regulations 2007

The Construction (Design and Management) Regulations (CDM Regulations) require active planning, coordination and management of the building works, including the electrical installation, to ensure that hazards associated with the construction, maintenance and perhaps demolition of the installation are given due consideration, as well as provision for safety in normal use.

Regulation 20 (2) requires a record known as 'the health and safety file' to be prepared, reviewed and updated during the construction process. This file should be passed to the client on completion of the construction work.

Regulation 17 (3) requires that reasonable steps are taken to ensure that once the construction phase has been completed the information in the health and safety file remains available for inspection by any person who might need it to comply with any relevant statutory provisions. It also requires that the file is revised and updated as often as may be appropriate to incorporate any relevant new information.

1.2.6 The Dangerous Substances and Explosive Atmospheres Regulations 2002

These Regulations place a duty on the employer to assess the risks arising from dangerous substances in the workplace. This includes the presence of flammable atmospheres and potential ignition sources. If such conditions cannot be eliminated, the regulations require the selection of electrical equipment that has protection to avoid it becoming a source of ignition appropriate for the areas in which it is to be used.

1.2.7 The Petroleum (Consolidation) Act 1928

This Act has particular application to any place where petroleum spirit is kept or handled.

Guidance on the design, construction, modification, maintenance and decommissioning of filling stations is published by the Association for Petroleum and Explosives Administration (APEA) and the Energy Institute, ISBN 0 85293 419 X. Copies of this document can be obtained from Portland Customer Services, Commerce Way, Colchester, CO2 8HP. Tel. 01206 796351.

Guidance on the safe use of petroleum is available on the Health and Safety Executive's website: www.hse.gov.uk/fireandexplosion/petroleum

1.2.8 The Cinematograph (Safety) Regulations 1955

These Regulations apply to premises used for cinematography exhibitions.

1.2.9 The Building Regulations 2000 (England and Wales)

Approved Document B 2006 was issued by the Department for Communities and Local Government (CLG) to provide practical guidance on meeting the requirements of Schedule 1 (means of escape) and Regulation 7 (materials and workmanship) of the Building Regulations.

Approved Document B consists of two parts as follows:

▶ Volume 1 – Dwellinghouses
▶ Volume 2 – Buildings other than dwellinghouses (amended 2007)

See Appendices B, C and D of this Guidance Note for further information.

1.2.10 The Building (Scotland) Regulations 2004

Compliance with the Scottish Building Regulations can be achieved by compliance with the mandatory Building Standards and associated guidance given in the two Scottish Building Standards Technical Handbooks (Domestic and Non-Domestic).

See Appendix E of this Guidance Note for further information.

1.2.11 The Building Regulations (Northern Ireland) 2000 as amended

The Department of Finance and Personnel has produced *Technical Booklet E:2005* and also refers to other deemed-to-satisfy publications, such as British Standards, to support compliance with Part E (Fire safety) of the Building Regulations (Northern Ireland) 2000.

1.2.12 The Health and Safety (Safety Signs and Signals) Regulations 1996

These Regulations require employers to use a safety sign where there is a significant risk to health that cannot be controlled or avoided by other means. The HSE guide, *Safety Signs and Signals* (L64), describes in detail a range of signs including Emergency Escape and Fire Safety Signs.

1.2.13 The Regulatory Reform (Fire Safety) Order 2005

This is probably the single most influential statutory instrument in the field of protection against fire. It has a direct influence on other pieces of primary and secondary statutory legislation, requiring modifications to and in some cases partial or full revocation of the requirements therein. It replaces fire certification under the Fire Precautions Act 1971 with a general duty to ensure, so far as is reasonably practicable, the safety of employees and a general duty, in relation to non-employees, to take such fire precautions as may reasonably be required to ensure that premises are safe. It also requires a risk assessment to be carried out and regularly updated.

The Regulatory Reform (Fire Safety) Order 2005 (henceforth referred to as the Order) came into effect fully in October 2006 and affects over 70 pieces of fire safety law, not all of which fall within the scope of this publication.

The Order applies to non-domestic premises in England and Wales, including any communal areas of blocks of flats or houses in multiple occupation, and places legal obligations and responsibilities upon the **responsible person** who is defined in article 3 of the Order as follows:

3. Meaning of 'responsible person'

In this Order 'responsible person' means –

(a) in relation to a workplace, the employer, if the workplace is to any extent under his control;

(b) in relation to any premises not falling within paragraph (a) –

   (i) the person who has control of the premises (as occupier or otherwise) in connection with the carrying on by him of a trade, business or other undertaking (for profit or not); or

   (ii) the owner, where the person in control of the premises does not have control in connection with the carrying on by that person of a trade, business or other undertaking.

Part 2 of the Order (Fire Safety Duties) requires the responsible person to take the necessary general fire precautions to ensure, so far as is reasonably practicable, the safety of employees and other persons within the premises for which they are responsible. The responsible person must carry out a fire safety risk assessment on the premises to identify what general fire precautions are required for the above to be achieved. The risk assessment must be regularly reviewed by the responsible person to keep it up to date.

The significant findings of the risk assessment including any actions that have been taken or which will be taken by the responsible person and details of any persons identified as being especially at risk must be recorded as soon as practicable afterwards where:

▶ the responsible person employs five or more persons, or
▶ the premises to which the assessment relates are subject to a licensing arrangement, or
▶ the premises are subject to an alterations notice.

Article 23 requires employees to:

a take reasonable care for the safety of themselves and of other persons who may be affected by their acts or omissions at work;

b cooperate with their employer as regards any duty or requirement imposed by or under any provision of the Order, so far as is necessary to enable that duty or requirement to be performed or complied with; and

c inform their employer or any other employee with specific responsibility for the safety of fellow employees –

   i of any work situation which they would reasonably consider represented a serious and immediate danger to safety; and

   ii of any matter which they would reasonably consider represented a shortcoming in the employer's protection arrangements for safety,

in so far as that situation or matter either affects the safety of the employee or arises out of or in connection with their activities at work, and has not previously been reported to the employer.

Article 37 relates to fire-fighters' switches for luminous tube signs etc. and is a topic dealt with in Guidance Note 2: *Isolation & Switching*. GN2

Reference should be made to the various schedules to the Order, which are as follows:

Schedule 1   This consists of four parts. Parts 1 and 2 relate to risk assessment, Part 3 relates to principles of prevention and Part 4 relates to measures to be taken in respect of dangerous substances.

Schedule 2   Amendments of primary legislation

Schedule 3   Amendments of subordinate legislation

Schedule 4   Repeals

Schedule 5   Revocations

It should be noted that in Scotland the Fire (Scotland) Act 2005 provides similar requirements for fire safety duties to those of the Regulatory Reform (Fire Safety) Order 2005 applicable to England and Wales.

1.2.14 Other Regulations

Electrical installations are subject to many local authority bylaws and conditions of licence. The following may be mentioned:

► places licensed for public entertainment, music, dancing, etc.

► large buildings, office blocks and the like, having limited access and escape points

► buildings to which the public have access, covered shopping centres and the like, hotels and boarding-houses.

Section 20 of the London Building Acts (Amendment) Act 1939 (as amended primarily by the Building (Inner London) Regulations 1985) is principally concerned with the danger arising from fire within certain classes of building which, by reason of height, cubic capacity and/or use, necessitate special consideration. The types of building coming within these categories are defined under Section 20 of the amended 1939 Act.

Note: These regulations have been subject to some repeals under the Regulatory Reform (Fire Safety) Order 2005. See Schedule 4 of the Order for full details.

The designer of the electrical installation in any building falling within these descriptions should make the earliest possible approach to the Local Authority and the Local Fire Authority to establish their requirements, obtain advice and get written agreement to the proposed arrangements.

Overview of the Wiring Regulations

2

This Guidance Note follows, as far as possible, the sequence in which matters relating to protection against fire appear within the 17th Edition. This approach has been adopted in order to make for its easy use in conjunction with the 17th Edition itself.

Chapter 13 of BS 7671 prescribes the fundamental principles for safety and includes a number of regulations which are directly concerned with the fire risk associated with electrical installations.

Chap 13

In particular, Regulation 131.3.1 deals with the risk of ignition of flammable materials due to high temperature or electric arc, and Regulation 131.3.2 deals with the effects of heat or thermal radiation emitted by electrical equipment. Other chapters set out in greater detail the methods and practices which are regarded as meeting the requirements of Chapter 13.

131.3.1
131.3.2

For example, Chapter 42, 'Protection against thermal effects', has requirements for:

Chap 42

▶ protection against fire caused by electrical equipment,
▶ precautions where particular risks of fire exist, and
▶ protection against burns.

421
422
423

Section 421, 'Protection against fire caused by electrical equipment', requires measures to be taken to prevent electrical equipment from presenting a fire hazard to materials that are, or could conceivably be, found in close proximity to such equipment.

Section 421

Section 422, 'Precautions where particular risks of fire exist', contains a number of specific requirements for wiring systems and switchgear/controlgear located within escape routes. Concern is often expressed over the impairment of escape routes by the smoke and fumes generated by fire. This is an important but very complex issue, and the risk from electrotechnical products should not be considered separately but as a part of the overall fire hazard assessment.

Section 422

Section 422 also contains measures in addition to those of Section 421 to be applied in installations or locations therein in buildings mainly constructed of flammable materials such as wood.

422.4

Section 423, 'Protection against burns', states maximum temperature limits not to be exceeded on the external surfaces of fixed electrical equipment in normal use.

Section 423

Chapter 43 prescribes requirements which limit the temperatures of live conductors and their insulation to prevent damage to the cable under overload and fault conditions.

Chap 43

Regulation 522.1.1 requires a wiring system to be installed such that it is suitable for both the highest and lowest ambient temperatures likely to be experienced. Further, it

522.1

Table 52.1
Table 43.1

requires a cable to be selected and installed such that the limiting temperature of its insulation as stated in Table 52.1, and its limiting temperature under fault conditions given in Table 43.1, will not be exceeded.

Section 527

Section 527 prescribes requirements for wiring systems to be installed to minimise the spread of fire, which are described in Chapter 6 of this Guidance Note.

Section 532

Section 532 gives requirements for devices being selected to provide protection against fire.

Chap 54
GN6

Chapter 54 has the same intent as Chapter 43 as regards protective conductors under earth fault conditions. Both these chapters are the subject of Guidance Note 6: *Protection Against Overcurrent*.

Chap 55
554.4.4
554.5.1
559.5.1
559.11.4

Chapter 55 gives requirements for items of other equipment. Within this chapter requirements pertinent to protection against fire are now given for heating conductors and cables; electric surface heating systems; low voltage luminaires and lighting installations; and extra-low voltage lighting installations.

Chap 56

Chapter 56 is of relevance to this publication as it gives requirements for safety services.

Section 611

Section 611 contains an inspection requirement for the presence and indeed effectiveness of fire barriers, seals and similar measures provided for protection against thermal effects.

621.2

Section 621 contains specific requirements for periodic inspection to provide for the safety of persons and livestock with regard to electric shock and burns, and protection against damage to the property caused by fire and heat resulting from an installation defect.

705.422.7

Section 705, 'Agricultural and horticultural premises', includes a requirement for the installation of an RCD having a rated residual operating current not exceeding 300 mA for protection against fire.

Section 711

Section 711 contains a number of requirements relating to the installation of luminaires and other items of equipment having high surface temperatures installed in installations for exhibitions, shows and stands.

740.422.3.7

Section 740 requires a manually resettable thermal cut-out to be fitted to motors which are automatically and/or remotely controlled and unsupervised where installed in amusement parks, circuses, fairgrounds and similar temporary installations.

Section 753

Section 753 deals with floor and ceiling heating systems and contains a number of requirements relating to protection against burns and overheating.

Where consideration of the effects of smoke, fumes, fire propagation and/or circuit integrity is necessary, cables which have characteristics classified in standard fire tests can be used. The cable standards are:

BS 6724

▶ BS 6724:1997 (2008) – *Electric cables. Thermosetting insulated, armoured cables for voltages of 600/1000 V and 1900/3300 V, having low emission of smoke and corrosive gases when affected by fire*

▶ BS 7211:1998 – *Electric cables. Thermosetting insulated, non-armoured cables for voltages up to and including 450/750 V, for electric power, lighting and internal wiring, and having low emission of smoke and corrosive gases when affected by fire*   BS 7211

▶ BS 7629-1:2008 – *Electric cables. Specification for 300/500 V fire resistant screened cables having low emission of smoke and corrosive gases when affected by fire. Multicore and multipair cables*   BS 7629-1

▶ BS 7846:2000 – *Electric cables. 600/1000 V armoured fire-resistant cables having thermosetting insulation and low emission of smoke and corrosive gases when affected by fire*   BS 7846

Note: Many cable standards have been amended as from 1 April 2004. See standards listing in Appendix A.

Cables clipped direct may also be mineral insulated to BS EN 60702-1:2002. These may be either bare or, where a covering is required, it should be designated as having 'low emission of smoke and corrosive gases when affected by fire'.   BS EN 60702-1

General guidance on the selection of cables is given in BS 7540 series (2005) – *Electric cables. Guide to use for cables with a rated voltage not exceeding 450/750 V.*   BS 7540

Generally applicable requirements 3

3.1 General

The primary purpose of BS 7671 is to ensure that persons, property and livestock are protected against hazards arising from the electrical installation. The particular risks identified are:

- shock currents
- excessive temperatures which may cause injurious effects such as burns and fire
- the ignition of a potentially explosive atmosphere
- undervoltages, overvoltages and electromagnetic influences which are potentially damaging or injurious
- injury from mechanical movement of electrically actuated equipment
- power supply interruptions/interruption of safety services
- arcing or burning that is likely to cause blinding effects, the production of excessive pressure and/or toxic gases.

131.1

Of the above, excessive temperatures, ignition of potentially explosive atmospheres, and arcing and burning effects are considered in this publication.

Persons, livestock, fixed equipment and fixed materials, whether part of the fabric of the building or the contents thereof, should be protected against harmful thermal effects which may result in combustion, ignition or degradation, the risk of burns, or the impairment of the safe functioning of equipment and hence the safe use of the installation for its intended purpose.

131.3.2

Harmful effects can occur as a result of:

420.3

- heat accumulation
- radiated heat
- hot components or items of equipment
- failure of protective devices, switchgear, thermostats, temperature limiting devices
- inadequate provision or failure of sealing arrangements where wiring systems penetrate the fabric of a building or within wiring systems themselves (where required)
- overcurrent
- insulation faults
- arcing, sparking or high temperature particles
- harmonic currents
- lightning.

3.2 Fixed equipment

Section 421   Although there is a tendency for fires of unexplained cause to be blamed upon faulty electrical installations and electrical appliances, many fires can be correctly attributed to faults within, or incorrect selection and use of, either the fixed installation or electrical equipment. Neither the electrical installation designer nor the installer has any control over the selection and use of portable equipment and so they are not considered within this guidance note. Chapter 42 is concerned solely with fixed equipment. Other sections of BS 7671 contain measures to minimise the risk of fire specific to particular types of equipment and these are summarised below.

3.2.1 Heating cables

554.4.1   Where it is necessary for a heating cable to either pass through or be in close proximity to any material which presents a fire hazard, it should be enclosed in material having

BS 476-12   the ignitability characteristic 'P' as specified in BS 476-12:1991 *Method of test for ignitability of products by direct flame impingement*.

554.4.4   Furthermore, the loading of a floor-warming cable should be so limited that the manufacturer's maximum stated operating temperature is not exceeded in normal operation, due consideration being given to other factors which may further limit the maximum operating temperature, such as terminations, accessories and materials with which they are likely to be in contact.

554.5.1   Electric surface heating systems should be installed in accordance with BS 6351-2:1983
BS 6351-2   (2007) *Guide to the design of electric surface heating systems* and BS 6351-3:1983
BS 6351-3   (2007) *Code of practice for the installation, testing and maintenance of electric surface heating systems*.

Section 753   Reference should also be made to Section 753, which contains requirements for floor and ceiling heating systems.

3.2.2 Luminaires

559.5.1   The thermal effects of radiant and convected energy on the surroundings should always be considered when luminaires are selected and installed. Factors to be considered are: the fire resistance of materials in close proximity to, or likely to be affected by, the heat produced by the luminaire; confirmation that recommended minimum clearances between luminaires and combustible materials have been observed (see

422.3.1   also requirements of Regulations 422.3.1 and 422.4.2, which are discussed further in
422.4.2   sections 4.3 and 4.4 respectively of this publication); and the use of lamps which do not exceed the maximum rating for the luminaire.

3.2.3 Extra-low voltage lighting

414.4.5   Regulation 414.4.5 makes it clear that basic protection such as insulation is not required in normal dry conditions for:

▶ SELV circuits having nominal voltage not exceeding 25 V a.c. or 60 V d.c.
▶ PELV circuits having nominal voltage not exceeding 25 V a.c. or 60 V d.c. where the exposed-conductive-parts and/or live parts are connected by a protective conductor to the main earthing terminal.

559.11.4.1   In the case of extra-low voltage lighting, and both live conductors are uninsulated,
BS EN 60598-2-23   then the lighting system should comply with BS EN 60598-2-23, or they should
559.11.4.2   be protected by a device meeting all the requirements of Regulation 559.11.4.2 as described below:

- ▶ continuously monitors the power demand of the connected luminaires
- ▶ capable of automatically disconnecting the supply circuit within 0.3 second in the event of a short-circuit or other failure which results in a power increase of more than 60 W
- ▶ provides automatic disconnection when the circuit is operating with reduced power such as may occur as a result of lamp failure, gating control or similar, or any failure causing a power increase of more than 60 W
- ▶ provides automatic disconnection on connection of the supply circuit in the event of any failure which results in a power increase of more than 60 W
- ▶ the device should fail-safe.

3.2.4 Symbols used in connection with luminaires, controlgear for luminaires and their installation

The following symbols, which may be found on luminaires and controlgear for luminaires or their packaging data sheets, are of relevance to the subject of protection against fire.

Table 55.2

Note: The symbols relating to luminaires are to be changed as a result of international (IEC) work and the release of a new version of BS EN 60598-1. Products may therefore be sold with the old and/or new symbols displayed on them or their packaging.

The following symbols come from the 2004 edition of BS EN 60598-1:

BS EN 60598-1:2004

 Luminaire with limited surface temperature

 Luminaire suitable for direct mounting on normally flammable surfaces

 Luminaire suitable for direct mounting on non-combustible surfaces only

 Luminaire suitable for mounting in/on normally flammable surfaces where a thermally insulating material may cover the luminaire

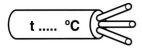 Use of heat-resistant supply cables, interconnecting cables or external wiring

 Rated maximum ambient temperature

 Warning against the use of cool-beam lamps

 Minimum distance from lighted objects (m)

Transformer – short-circuit proof (inherently and non-inherently)

 'class P' thermally protected ballast/transformer

 Temperature declared thermally protected ballast/transformer, with a value equal to or below 130°C

BS EN 60598-1:2008 The following symbols come from the 2008 edition of BS EN 60598-1 and will replace the 'F mark' symbols shown above.

 Surface mounted luminaire not suitable for direct mounting on normally flammable surfaces (suitable only for mounting on non-combustible surfaces – surface mounting)

 Recessed luminaire not suitable for direct mounting on normally flamma-ble surfaces (suitable only for mounting on non-combustible surfaces – recessed mounting)

 Luminaire not suitable for covering with thermally insulating material.

3.3 Manufacturers' instructions

421.1
134.1.1

510.2

When electrical equipment is installed, any reasonable and relevant instructions of the equipment manufacturer should be followed. This is a reminder of the general requirement that electrical equipment be installed in accordance with any instructions provided by the manufacturer. Where an installation includes fixed equipment, the fact that the selection of the equipment may have been made by other than the installer does not absolve the installer from the responsibility of seeing that any installation instructions of the manufacturer are met.

3.4 No fire risk in normal use

421.2

The heat generated by fixed electrical equipment in normal use should not be capable of causing a fire or harmful thermal effects to adjacent fixed material. Further, the installation designer is required to exercise foresight and consider other material which is likely to be in proximity to such equipment.

Regulation 421.2 offers three installation methods for equipment which in normal operation has a surface temperature sufficient to cause a risk of fire or harmful effects to adjacent materials. In short, such equipment should be supported, enclosed or screened by items capable of withstanding the operating temperature, or be mounted in a manner that permits safe heat dissipation and gives adequate clearance from surrounding equipment or materials.

One or more of the methods must be adopted and as far as Regulation 421.2(iii) is concerned, Figure 3.1 gives guidance for equipment which in normal operation has a surface temperature exceeding 90 °C.

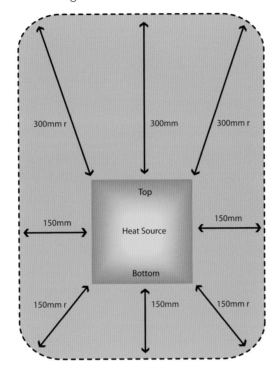

▼ **Figure 3.1**
Clearances required to achieve requirements of Regulation 421.2(iii)

By way of example, attention must be given to recessed or semi-recessed luminaires and luminaire controlgear mounted in ceiling voids to ensure that heat is properly dissipated and that any thermally insulating materials are not and cannot become so disposed as to restrict the cooling of the equipment.

This might be achieved by the use of a fully closed back 'fire-rated' type downlighter such as the one shown in Figure 3.2, which is particularly suitable for installation into ceiling/floor voids where access above the luminaire is not possible.

▼ **Figure 3.2**
Fire-rated downlighter

Where access to the space above the luminaire can be achieved, such as for example in a domestic loftspace, a proprietary insulation support box such as the one shown in Figure 3.3 may suffice if installed over more open-backed design recessed luminaires.

▼ **Figure 3.3**
Insulation support box

BS EN 61184   Lampholders are one example of equipment having temperature ratings. BS EN 61184:1997 *Bayonet lampholders* gives guidance on the selection of lampholders for particular applications. However, because it is not possible to guarantee how an installation will be used, every B15 or B22 lampholder must be T2 temperature rated to comply with BS 7671. Where lampholders other than B15 or B22 are installed, their rated operating temperature must be suitable for the application.

559.6.1.7

3.5 Arcs, sparks and hot particles

421.3   The effects of arcs or the emission of high temperature particles must be guarded against. In some equipment, e.g. air circuit-breakers or semi-enclosed fuses, emissions may be produced during fault clearance and the enclosures of such equipment must comply with the appropriate British Standard, be totally enclosed in arc-resistant material, or be screened by arc-resistant material, or be mounted so as to allow for the safe extinction of the emissions at a sufficient distance from the material upon which the emissions could have a harmful effect.

511

By its application to any item of fixed equipment, BS 7671 also applies to permanently installed electric welding sets. Electric arc welding and similar (fixed or portable) is subject to Regulations 7, 8 and 14 of the Electricity at Work Regulations 1989. Regulation 7 requires insulation of live conductors, Regulation 8 requires earthing and Regulation 14 specifies requirements for work on or near live conductors.

EWR 1989

3.6 Focusing and concentration of heat

Equipment that focuses heat, such as radiant heaters and high intensity luminaires, must be positioned so that the building structure or other materials are not subjected to excessive temperatures.

421.4

The manufacturers' instructions must be followed with respect to spacing from walls (w), floor (f) and ceiling (c), and the angle of inclination θ of the heater (Figure 3.4). Regulation 422.3.1 states the **minimum** distances that spotlights and projectors must be installed from combustible materials; see section 4.3.1 of this publication.

422.3.1

Downlighters fitted with aluminised lamps project both the light and heat generated away from the light fitting. This heat can be considerable, therefore care should be taken to ensure that luminaires fitted with such lamps are not so placed that, for example, a door can be left open sufficiently close below for a risk of fire to exist.

See also section 3.2 of this publication.

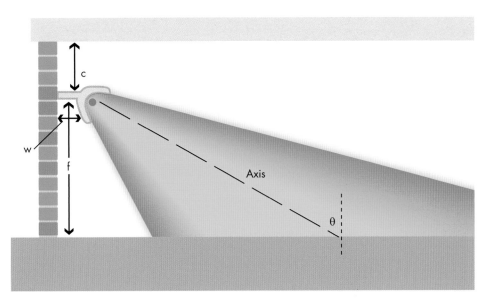

▼ **Figure 3.4**
Focusing of heat from a heater or luminaire

3.7 Equipment containing flammable liquid

Special precautions are necessary for flammable dielectric liquids, as fires involving these liquids are life-threatening within a few seconds of ignition. In the event of spillage, the object is to limit the spread of the liquid and the exposed surface area, thus limiting the size of the fire and the danger to persons and property.

421.5

Where a large quantity of liquid is involved, more than one escape route is necessary, in the same way as for a large switchroom. It is important to ensure that persons have alternative routes so that the main seat of a fire may be avoided.

The 'single location' referred to in Regulation 421.5 is the location containing all the flammable liquid which may be involved at the outset of an emergency. The amount of flammable liquid in an area must therefore be established, whether contained in one item of equipment or in a number of separate items.

The options available to the designer will depend on a number of considerations, for example, whether a single item of equipment is involved or a number of items and whether the location is indoors or outdoors. The options include:

▶ reducing the risk by partitioning the location with fire doors and sills
▶ providing bunds or kerbs around items of equipment or, for larger items of equipment, a retention pit filled with pebbles or granite chips (the net capacity of the bund or retention pit when filled with pebbles or chips should exceed the oil capacity of the equipment by at least 10 per cent)
▶ providing a drainpit and flame arrestor
▶ provision of automatic fire venting and/or automatic fire suppression or foam inlets and integration with the automatic fire detection and alarm system of the building, where appropriate
▶ ramped floors
▶ use of an outdoor location
▶ blast walls between large items.

▼ **Figure 3.5**   An oil-filled transformer located in a plant room, which incorporates many of the measures described above: the location is fitted with a fire door, the floor falls to a central drainage point leading to a drainpit (grille just visible to left of picture) and the walls have been constructed to contain any blast in the event of explosion

In some circumstances, it may be appropriate to provide explosion venting for interior locations, and environmental considerations should always involve discussions with public health engineers concerning the provision of an oil interceptor to prevent contamination of the sewers.

When carrying out inspections of older installations the measures provided should be carefully examined in the light of modern practice. In particular, it may be that over a period of time, seepage and occasional spillage has caused a wooden floor to be saturated with oil and thus become an additional fire risk.

Adequate general lighting and emergency lighting should be provided where appropriate (particularly on escape routes).

First-aid equipment should be provided where appropriate.

3.8 Enclosures

Where enclosures are constructed on site, reference must be made to the appropriate   421.6
product standard for the necessary resistance to heat and fire. In the absence of such
a standard, maximum likely temperatures of the enclosure must be determined and
enclosures selected that are well able to withstand these temperatures.

BS 476 *Fire tests on building materials and structures* (Part 4 – *Non-combustibility test for materials*, Part 12 – *Method of test for ignitability of products by direct flame impingement*) provides type tests for materials.

Where equipment is installed in an enclosure or enclosed space, e.g. ballasts or other controlgear within a ceiling, adequate arrangements must exist for safely dissipating heat generated.

3.9 Terminations

Regulation 421.7 is the general requirement for the enclosure of terminations of   421.7
live conductors and joints. Reference is made to Regulation 526.5 where particular   526.5
requirements are given, including also the enclosure of terminations of PEN conductors.
A further important requirement is to provide an enclosure which will protect the joint   526.7
or connection against the environment.

Poor terminations and connections are cited as a frequent cause of fire, and close   526.1 to 526.9
attention is required by the designer, the installer and the person responsible for the   611.3
inspection and testing to all aspects of the subject.

For this reason, Regulation 526.5 requires that all terminations and joints at whatever   526.5
voltage, including ELV, shall be made within a suitable enclosure. Because there is
always a risk of overheating and consequent fire at a joint or termination, these must
be enclosed and the enclosure where not incorporated in suitable equipment or
accessory must meet the specified fire resistance requirements. This applies equally
to ELV connections to luminaires and similar equipment. The relatively high currents
of ELV equipment mean particular care has to be taken with joints and terminations.
When compression joints are used, crimping tool, lug and cable must be compatible.

Other matters which should be considered under this heading are:

i   dirty or misaligned equipment contacts which may give rise to local heating, and

ii   loose or inadequate cable supports which may place mechanical stresses on connections, causing overheating.

One method, which could be considered when inspecting electrical installations for signs of overheating, is the use of thermal imaging. This is not required by BS 7671 and would therefore be in addition to the inspections and tests stated in Part 6 of BS 7671. Thermal imaging of exposed live parts of electrical installations must comply with the requirements of Regulation 14 of the Electricity at Work Regulations 1989. Regulation 14 prohibits a person working on or near any exposed bare live conductor unless it is unreasonable in all circumstances for it to be dead, and it is reasonable in all circumstances for the person to be at work on or near it while it is live and suitable precautions are taken to prevent injury.

The use of thermographic surveying is also discussed in Guidance Note 3: *Inspection & Testing*.

Precautions where particular risks of fire exist

4

4.1 General

BS 7671 considers particular risks of fire to exist in escape routes from a building; locations having risks of fire due to the nature of materials being processed or stored therein; where the fabric of the building has been constructed mainly of combustible materials; where the structure of the building aids the propagation of fire; and where the building itself or indeed its contents are of some particular significance.

Section 422

Generally, in such locations, only equipment necessary for the intended use of the location should be installed. Exceptions to this requirement are given in Regulation 422.3.5 and are considered later.

422.1.1

Any installed equipment should be so constructed that any normal or foreseeable temperature rise, even if present under fault conditions, cannot result in a fire occurring. If necessary this requirement may be achieved by the application of additional protective measures.

Any thermal cut-out devices installed in a location covered by Section 422 should have a manual reset facility.

422.1.3

4.2 Requirements for emergency escape routes

Those parts of a building which constitute part of an emergency escape route are classified in Appendix 5. Regulation 422.2 makes it clear that no special requirements apply in conditions having the classification BD1, which is domestic premises and other places of low density occupation and of low to normal building height.

Appx 5
422.2

Regulation 422.2 does contain a number of specific requirements for locations classified as BD2 (multi-storey buildings such as offices), BD3 (buildings open to the public, such as shopping centres and places of public entertainment) and BD4 (high-rise buildings open to the public, such as hotels).

In buildings classified BD2, BD3 or BD4, wherever possible wiring systems should not encroach on an escape route and should in any case be as short as is possible. Further, only wiring systems having sufficient protection against mechanical damage likely to be encountered during evacuation in such a location may be installed within arm's reach.

422.2.1

All wiring systems employed, including both the cables themselves and the supporting structure provided, must be non-flame propagating. This can be achieved by the use of:

BS EN 50266 ▶ cables meeting the appropriate fire tests described in BS EN 50266

BS EN 61386-1 ▶ conduit systems meeting the fire tests described in BS EN 61386-1

BS EN 50085 ▶ trunking and ducting systems meeting fire tests described in BS EN 50085.

Any wiring within the escape route which supplies a safety service must, in the absence of a more strenuous requirement for building elements, have a resistance to fire of two hours.

A decent level of visibility in escape routes should be maintained for as long as possible. To achieve this, wiring in such locations should be of a type having limited smoke production characteristics. Where no more stringent requirement exists in a cable product standard, a value of 60 per cent light transmittance is recommended for BS EN 61034-2 cables tested to BS EN 61034-2.

Although not specifically required by BS 7671, it is highly recommended that only metallic clips or cable ties are employed where the failure of such a support could result in a danger of entanglement or otherwise impair the effectiveness of the escape BS 5839-1 route. BS 5839-1:2002 (incl. AMD 2 2008) specifically precludes the use of plastic cable clips, ties and trunking as the sole means of supporting cables forming part of the wiring of a fire alarm system, both as a means of maintaining circuit integrity and to minimise the hazard presented by unsecured cables under fire conditions. The preclusion on the use of plastic clips, ties and the like as the sole means of cable BS 5266-1 support also appears in clause 9.2.2 of BS 5266-1 (Emergency lighting).

422.2.2 Any switchgear or controlgear other than fire alarm call points installed in a building classified BD2, BD3 or BD4 shall be so installed that it is only accessible to authorised persons. Further, if such items are installed within an emergency escape route an additional enclosure made of non-combustible or not readily combustible material must be provided. It should be noted that this requirement does not extend to accessories Part 2 such as light switches (see Definitions).

422.2.3 With the exception of small capacitors likely to be found in equipment such as discharge lights, equipment containing flammable liquids should not be installed in escape routes.

BS 5266 Reference should be made to the relevant parts of the BS 5266 series which gives the requirements for emergency escape lighting.

4.3 Risks of fire due to the nature of processed or stored materials

Appx 5 Appendix 5 classifies the risks associated with the nature of processed or stored materials. Classifications relevant to protection against fire are:

▶ BE2 – a fire risk exists as a direct result of the manufacturing, processing or storage of materials. This would include premises such as barns, woodworking shops, industrial scale bakeries and paper mills.

▶ BE3 – an explosion risk exists from the processing or storage of low flashpoint materials. This would include petrochemical plants and hydrocarbon fuel storage facilities.

422.3 BS 7671 contains no specific requirements where an explosion risk exists, but rather BS EN 60079-14 refers to BS EN 60079-14:2008 *Explosive atmospheres. Electrical installation design, selection and erection.*

Regulation 422.3 contains a number of requirements (described below) to be applied in addition to the general requirements for installations in locations classified as BE2, where the manufacturing, processing or storage of combustible materials and the accumulation of materials such as dust and fibres constitutes a risk of fire but not explosion.

422.3

4.3.1 Luminaires

Except where otherwise recommended by the manufacturer, spotlights and projectors should be installed at the following minimum distances from combustible materials:

422.3.1

▶ rating up to 100 W – 0.5 m
▶ rating over 100 W up to 300 W – 0.8 m
▶ rating over 300 W up to 500 W – 1.0 m.

Only luminaires having enclosures providing a degree of protection of at least IP5X, and with limited surface temperatures in accordance with BS EN 60598-2-24, should be employed. Fittings manufactured to BS EN 60598-2-24 have horizontal surface temperatures limited to 90 °C under normal conditions and 115 °C under fault conditions.

422.3.8
BS EN 60598-2-24

Additionally, luminaires should be of a type that prevents components which are likely to run hot, such as lamps, from falling from the luminaires.

4.3.2 Enclosures

Enclosures of equipment such as heaters should not be able to attain higher surface temperatures than 90 °C under normal conditions, or 115 °C under fault conditions.

422.3.2

Where it is likely that dust or fibres could accumulate on an enclosure of electrical equipment in such a manner or quantity as to cause a fire hazard, precautions need to be taken to prevent said enclosure from exceeding the aforementioned temperatures. Good housekeeping is also necessary in such locations.

4.3.3 Switchgear and controlgear

Electrical equipment should be installed outside the location unless suitable for the location, or installed inside a suitable enclosure. A suitable degree of protection for such an enclosure would be IP4X or, if the presence of dust is to be reasonably expected, IP5X. The IP code system is covered by BS EN 60529:1992 (2004) *Specification for degrees of protection provided by enclosures (IP code)*. To summarise, the degree of protection provided by an enclosure is indicated by two numerals. The first indicates the degree of protection against solid bodies and the second indicates the degree of protection against liquids. Where a characteristic numeral is not required to be specified it can be replaced by the letter X. (Refer to Guidance Note 1, Appendix B for full details.) For IP5X, the first numeral 5 indicates protection against contact by wires or strips more than 1.0 mm thick, and dust-protected (dust may enter but not in an amount sufficient to interfere with satisfactory operation or impair safety).

422.3.3

BS EN 60529

GN1 Appx B

4.3.4 Cables, circuits and wiring systems

Cables not completely embedded in non-combustible material, such as plaster or concrete, or otherwise protected from fire shall meet the flame propagation characteristics as specified in BS EN 60332-1-2.

422.3.4

BS EN 60332-1-2

Conduit systems should be able to satisfy the test under fire conditions specified in BS EN 61386-1. Ducting and trunking systems should be able to satisfy the test under fire conditions specified in BS EN 50085.

BS EN 61386-1
BS EN 50085

BS EN 61537   Cable tray or cable ladder systems should be able to satisfy the test under fire conditions specified in BS EN 61537.

All such systems shall be non-flame propagating. Flame propagating systems may only be used where they are embedded in the building structure, e.g. screeded systems.

BS EN 50266   Where the risk of flame propagation is high, e.g. in long vertical runs or bunched cables, the cable must meet the flame propagation characteristics as specified in one of the categories of test in Parts 2.1 to 2.5 of BS EN 50266.

422.3.5   A wiring system which passes through, but is not intended for electrical supply within, the location should meet the requirements of Regulation 422.3.4 above, and have no joint or connection within the location unless it is placed in a suitable enclosure that does not adversely affect the flame propagation characteristics of the wiring system,

422.3.10   and be suitably protected against overcurrent as required by Regulation 422.3.10.

Bare live conductors should not be installed.

422.3.9   With the exceptions of mineral insulated cables complying with BS EN 60702-1,
BS EN 60702-1   powertrack systems complying with BS EN 61534, and busbar trunking systems
BS EN 61534   complying with BS EN 60439-2, wiring systems should be protected against insulation
BS EN 60439-2   faults to earth as follows:

i   In TN and TT systems, by an RCD with a rated residual operating current not exceeding 300 mA. Where a resistive fault in, for example, an overhead heating system employing heating film elements may cause a fire, 30 mA RCD protection will be required.

ii   In IT systems, by insulation monitoring devices with audible and visual signals. Adequate supervision is required to facilitate manual disconnection as soon as appropriate. In the event of a second fault, the disconnection time should be in accordance with the requirements of Regulation 411.6.4.

532.1   Where an RCD is used to meet the requirements of Regulation 422.3.9 as described above, the RCD should:

▶ be installed at the origin of the circuit for which it provides protection, and
▶ switch all live conductors, and
▶ have a rating not exceeding 300 mA.

422.3.10   All circuits supplying equipment within the location or which pass through the location must be protected against overload and faults by a device on the supply side of the location and placed outside the location.

422.3.11   All live parts of extra-low voltage circuits, regardless of the nominal voltage, must either:

▶ be contained in an enclosure having a degree of protection of at least IP2X or IPXXB, or
▶ have insulation capable of withstanding a test voltage of 500 V d.c. for 1 minute.

The use of PEN conductors is not permitted for circuits within the location, but is permitted for wiring passing through the location. It should be remembered at this point that the Electricity Safety, Quality and Continuity Regulations 2002 prohibit the use of PEN conductors in consumers' installations, with very few exceptions which are described in Regulation 543.4.2. PEN conductors are most likely to be forming part of either a public or private supply arrangement.

422.3.12
543.4.1
543.4.2

Every circuit, subject to the exceptions given in Regulation 537.1.2, should be provided with a means of isolation which disconnects all live conductors by means of a linked switch or linked circuit-breaker.

422.3.13

Any flexible cables and flexible cords employed shall be of heavy duty type having a voltage rating of not less than 450/750 V or be suitably protected against mechanical damage.

422.3.14

4.3.5 Motors

All motors which are automatically or remotely controlled, or which are not continuously supervised, should be protected against excessive temperature by a protective device with manual reset. Motors with star-delta starting must be protected against excessive temperature when connected in both star and delta.

422.3.7

It is also advisable to protect motors with slip-ring starters from being left with the resistance in the rotor. This type of rotor has a 3-phase winding, the ends of which are connected to three slip-rings on the rotor shaft. For this reason it is sometimes called a 'slip-ring' motor. This enables an external resistance to be added to the rotor circuit, which is used to:

a   limit the starting current
b   give a high starting torque
c   provide speed control.

At normal speed the slip-rings are short-circuited and the brush gear is lifted clear of the slip-rings to reduce wear.

4.3.6 Heating appliances

Any heating appliances installed in the location should be installed as fixed equipment, and heat storage appliances should be of a type which prevent the ignition of combustible materials and/or fibres by the heat storing core.

422.3.15
422.3.16

4.4   Combustible constructional materials

Regulation 422.4 has the following requirements where a building is mainly constructed of combustible materials such as wood (classified in Appendix 5 as CA2) and there is no explosion risk.

422.4

4.4.1 Equipment

Precautions should be taken so that electrical equipment does not pose an ignition hazard to walls, floors or ceilings to which it is in close proximity, by the adoption of appropriate design, installation methods and choice of electrical equipment.

422.4.1

Distribution boards and accessory boxes for switches, socket-outlets and the like that are installed into or on the surface of a wall made from combustible materials should meet the requirements of the relevant product standard for temperature rise of such an enclosure.

422.4.3

422.4.4 Where this is not the case, the equipment or accessory should be enclosed by non-flammable material of suitable thickness, taking into account the nature of the material being employed.

4.4.2 Luminaires

422.4.2 Except where an appropriate product standard states specific requirements for spacings from combustible materials, luminaires of the spotlight or projector type should be installed at the following minimum distances from combustible materials used to form the fabric of the building:

▶ rating up to 100 W – 0.5 m
▶ rating over 100 W up to 300 W – 0.8 m
▶ rating over 300 W up to 500 W – 1.0 m.

Luminaires should be of a construction that prevents components that are likely to run hot, such as lamps, from falling from the luminaires.

BS EN 60598-1 A luminaire that is marked ⟨F⟩ in accordance with BS EN 60598-1 is suitable for installation directly on a surface classified as normally flammable.

4.4.3 Cables and wiring systems

422.4.5
BS EN 60332-1-2 Any cables or cords installed in premises mainly constructed of combustible materials should meet the requirements of BS EN 60332-1-2:2004 *Tests on electric and optical fibre cables under fire conditions. Test for vertical flame propagation for a single insulated wire or cable. Procedure for 1 kW pre-mixed flame.*

422.4.6
BS EN 61386-1 Conduit systems should be in accordance with and meet the fire resistance requirements of BS EN 61386-1:2008 *Conduit systems for cable management. General requirements.*

BS EN 50085-1 Similarly, trunking systems should be in accordance with and meet the fire resistance requirements of BS EN 50085-1:2005 *Cable trunking systems and cable ducting systems for electrical installations. General requirements.*

4.5   Fire propagating structures

Appx 5 Buildings the shape, dimensions and layout of which facilitate the spread of fire are classified in Appendix 5 as CB2. Examples of such buildings would include high-rise buildings and buildings containing forced ventilation systems where a 'chimney effect' may exist having the detrimental effect of assisting a fire to remain burning.

422.5.1 Regulation 422.5.1 requires steps to be taken to ensure that the electrical installation in such buildings will not contribute to the spread of a fire.

Section 527 Particular attention should therefore be paid to reinstatement of structural integrity in relation to fire resistance where wiring systems penetrate the fabric of the building, and to the installation of fire-stopping materials within wiring systems where required by BS 7671.

Reference should also be made to section 6.2 of this publication which discusses the sealing of wiring system penetrations.

4.6 Locations of national, commercial, industrial or public significance

BS 7671:2008, the 17th Edition, now specifically mentions electrical installations in locations of national, commercial, industrial or public significance. This would include museums, national monuments, airports, railway stations, laboratories, computer and data storage centres, and archiving facilities.

Regulation 422.6 requires compliance with Regulation 422.1 and consideration of the following: 422.6

i   Installation of mineral insulated cables according to BS EN 60702   BS EN 60702
ii   Installation of cables with improved fire-resisting characteristics in case of a fire (i.e. shall meet the fire-resisting requirements of BS EN 50200 or BS 8491, as appropriate)   BS EN 50200 BS 8491
iii Installation of cables in non-combustible solid walls, ceilings and floors
iv   Installation of cables in areas with constructional partitions having a fire-resisting capability for a time of 30 minutes. In locations housing staircases and required for emergency escape purposes a 90-minute fire-resistance capability is required
v   Where i to iv are not practical, improved fire protection may be possible by the use of reactive fire protection systems (e.g. by a holistic fire engineering approach to the location).

Protection against burns

The requirements of Regulation 423.1 of BS 7671 apply only to protection against burns caused by contact with heated surfaces. Measures to prevent burns from heat radiation or arcing are covered by the requirements of Regulation 420.3.

423.1

420.3

Table 42.1 (reproduced here as Table 5.1) gives maximum acceptable surface temperatures for accessible parts of equipment within arm's reach during normal load conditions.

Table 42.1

Accessible part	Material of accessible surfaces	Maximum temperature (°C)
A hand-held part	Metallic	55
	Non-metallic	65
A part intended to be touched but not hand-held	Metallic	70
	Non-metallic	80
A part which need not be touched for normal operation	Metallic	80
	Non-metallic	90

▼ **Table 5.1**
Temperature limit under normal load conditions for an accessible part of equipment within arm's reach

Factors to be taken into account when using the table are whether the part is intended to be hand-held or touched in normal use, and of what materials the equipment is manufactured. Where the maximum temperatures prescribed in Table 42.1 are likely to be exceeded, albeit for a short period of time, the equipment in question should be fitted with guards or similar to prevent accidental contact.

Table 42.1 should not be applied to equipment for which a limiting temperature is specified in the relevant product standard.

423.1

It should be noted that mineral insulated cables exposed to touch are permitted to have a sheath temperature of 70 °C, corresponding to a metallic part intended to be touched but not hand-held. However, a cable having a conductor operating temperature of 90 °C may achieve a surface temperature approaching 80 °C in normal operation.

It must always be borne in mind that the temperatures in Table 42.1 are maximum values and that contact with any surface at or above 70 °C may cause a dangerous reflex action.

BS 4086:1966 (1995) *Recommendations for maximum surface temperatures of heated domestic equipment* provides technical considerations and recommended maximum temperatures for controls and working surfaces of heated domestic equipment. BSI PD 6504:1983 *Medical information on human reaction to skin contact with hot surfaces* provides information prepared by medical experts on human reaction to contact with heated surfaces. Both provide good guidance in determining 'safe' surface temperatures.

BS 4086

PD 6504

Even if the equipment complies with its standard as to surface temperature, consideration must still be given to the risk of burns, particularly when equipment is to be installed in locations to be used by the very young or infirm where additional precautions may be necessary, such as guards over heaters.

554.2.1   Regulation 554.2.1 requires every heater of liquid or other substance to incorporate, or to be provided with, an automatic device to prevent a dangerous temperature rise of the substance being heated. This requirement would apply, for example, to immersion heater elements heating the water stored in cylinders for domestic premises.

Measures to minimise the spread of fire

6

6.1 Precautions in a fire-segregated compartment

Section 527

Section 527 contains requirements regarding the selection of appropriate materials and the erection of the installation specifically aimed at minimising the spread of fire.

527.1.1

In all cases, a wiring system should be installed such that it does not detrimentally affect the structural integrity or fire safety of the building.

527.1.2

In installations where no particular extra risks of fire exist, standard thermoplastic insulated cables and standard flexible cables that comply with the requirements of BS EN 60332-1-2:2004 may be installed without the need for any other precautionary measures. However, in situations where a particular risk is identified within an installation (see Section 422 of BS 7671 and Chapter 4 of this publication) only cables that meet the flame propagation requirements given in the relevant part of the BS EN 50266 series *Common test methods for cables under fire conditions. Test for vertical flame spread of vertically-mounted bunched wires or cables* should be employed without additional protective measures being applied.

527.1.3

BS EN 60332-1-2

BS EN 50266

Standard cables as described earlier above may be employed in such cases if suitably enclosed for example in a metallic conduit, trunking or similar, or if installed in the building fabric such as being chased-in and plastered over.

527.1.5

The use of cables not complying with the flame propagation requirements of BS EN 60332-1-2, also known as the 1 kW Bunsen burner test, should be limited to only short lengths making the connection between appliances and the fixed installation. Such cables should not be installed such that they pass through more than one fire compartment.

527.1.4
BS EN 60332-1-2

Elements of a wiring system other than cables should meet the flame propagation requirements of the appropriate product standard as shown below:

527.1.5

▶ trunking and ducting – BS EN 50085
▶ conduit – BS EN 50086
▶ busbar trunking – BS EN 60439-2
▶ powertrack – BS EN 61534
▶ tray – BS EN 61537.

BS EN 50085
BS EN 50086
BS EN 60439-2
BS EN 61534
BS EN 61537

Where this is the case, or where components meeting similar flame propagation requirements in their appropriate product standard are used, no further precautions need be taken.

527.1.6

527.1.5 Where this is not the case, but the above components comply in all other ways with the appropriate product standard, they may only be employed if they are completely enclosed by non-combustible building materials.

6.2 Sealing of wiring system penetrations

6.2.1 External sealing arrangements

It is accepted that the passing of component parts of an installation (such as cables, tray and trunking) through ceilings, floors, walls and the like is unavoidable. However,

527.2.1 in order to reduce the possibility of spread of fire and the products of combustion to a minimum, it is important that any opening remaining around such parts where they penetrate an element of the building structure is made good to at least the same standard as that which existed prior to the penetration being made (see Figure 6.1).

527.2.2
527.2.3 It is also necessary to provide temporary sealing arrangements, as and where required, throughout the construction of an installation, and, wherever any sealing of penetrations around wiring systems is disturbed whilst work is being carried out, it should be reinstated as soon as possible.

527.2.7 Any sealing arrangement being used to meet the requirements of Regulations 527.2.1 and 527.2.2 as described above should:

▶ be capable of resisting external influences to the same degree as the wiring system component with which it is being used, and

▶ be equally resistant to the products of combustion and to penetration by water as the element of building fabric which has been penetrated, and

▶ be compatible with the material of the wiring system with which it is in contact such that neither is degraded, and

▶ permit sufficient thermal movement (through expansion and contraction) of the wiring system such that the sealing remains effective, and

▶ be sufficiently robust such that it can withstand the stresses likely to occur as a result of damage to the supports of the wiring system caused by fire. This condition is considered to be met where cable cleats, clips or other forms of support are installed within 750 mm of the seal and are capable of withstanding the strain following any such collapse.

▼ **Figure 6.1**
External and internal sealing arrangements

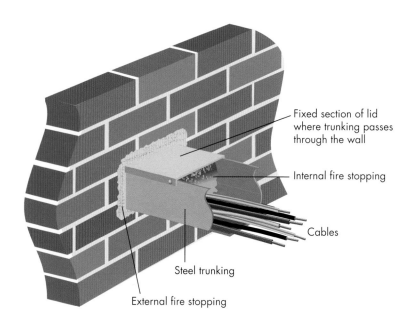

Fixed section of lid where trunking passes through the wall

Internal fire stopping

Cables

Steel trunking

External fire stopping

Furthermore, any such sealing arrangement and the wiring system itself unless fully resistant to moisture should be protected from dripping water which might travel along the wiring system to the seal or otherwise allow water to collect on or around the seal.

6.2.2 Internal sealing arrangements

With the exception of conduit, trunking or ducting systems classified as non-flame propagating in accordance with the relevant product standard, and having an internal cross-sectional area not exceeding 710 mm², any wiring system passing through an element of the building fabric having a specified fire resistance should be internally sealed to provide the same degree of fire resistance as the element of the building fabric being penetrated, in addition to the external sealing requirements described above. See Figure 6.1.

527.2.6

527.2.4

In practical terms this means that standard conduit of diameter 32 mm or less, or trunking or ducting not exceeding dimensions 25 x 25 mm or similar, will not require internal sealing.

It should be remembered that Regulation 611.3 (vii) calls for inspection, during the installation process and on completion of work, to verify the presence of fire barriers, seals and other measures provided for protection against thermal effects.

611.3

See also Appendices D and E of this publication for requirements relating to fire-stopping in the Building Regulations of England and Wales and of Scotland, respectively.

For installations in Northern Ireland reference should be made to the Northern Ireland Building Regulations website: www.buildingregulationsni.gov.uk

Safety services

7

7.1 General

The requirements of Chapter 56 are applicable to supplies to services which need to remain functional during a fire. This would include services such as booster pumps for sprinkler systems and risers; CO detection/alarm systems; emergency lighting; fire detection/alarm systems; fire evacuation systems; fire services communication systems; lifts designated for use by the fire service; industrial safety systems; life-critical medical systems; and smoke extraction equipment. It is not applicable to installations in hazardous areas, for which reference should be made to BS EN 60079-14, or to electrical standby systems.

560.1

BS EN 60079-14

It is worth remembering at this point that any wiring connecting a self-contained emergency luminaire to the main supply is not considered to be an emergency lighting circuit.

7.2 Classification and changeover times

Electrical safety service supplies may be:

560.4.1

▶ non-automatic, in which case the service is activated manually by an operator, or
▶ automatic, where operation does not require operator input.

Regulation 560.4.1 classifies automatic supplies by the maximum changeover time as follows:

▶ no-break – where a continuous supply is ensured subject to specified tolerances on, for example, voltage and frequency
▶ very short break – supply is available within 0.15 second
▶ short break – supply is available between 0.15 and 0.5 seconds
▶ lighting break – supply is available between 0.5 and 5 seconds
▶ medium break – supply is available between 5 and 15 seconds
▶ long break – supply is available in a time exceeding 15 seconds.

The response time used as the basis of any design should be in accordance with the requirements of the appropriate British Standard for the service in question. However, where no such standard exists it will be necessary to carry out a risk assessment to determine an acceptable changeover time.

560.4.2

Safety services can be expected to be called upon at all material times, that is when the premises in question are likely to be occupied by persons and/or livestock and hence those occupants may be at risk as a result of fire and/or the failure of the normal supply arrangements.

560.5.1

560.5.2   Where a safety service is required to remain fully operational under fire conditions, the safety source of supply (as distinct from the normal supply) must be maintainable for a sufficient duration to allow the safety service being supplied to perform its function correctly. Consideration must therefore be given to the requirements of appropriate, more specific British Standards. Taking emergency lighting by way of example, BS 5266-7:1999 (BS EN 1838:1999) specifies a minimum duration for emergency escape lighting of 1 hour (see clauses 4.2.5 and 4.3.5) whilst BS 5266-1:2005 specifies a minimum duration of 3 hours in premises where persons are likely to sleep who may be unfamiliar with the layout of the premises or may be infirm, such as hospitals, nursing homes, schools, colleges and hotels. Any specification for such emergency lighting would have to take these durations into account.

BS 5266-7 (BS EN 1838)
BS 5266-1

Furthermore, any installed equipment should either by its construction or as a result of the way in which it has been installed have a sufficient degree of fire resistance to enable it to remain operational for the prerequisite time.

560.5.3   It is preferred that any measures providing fault protection do not result in automatic disconnection under first fault conditions. For example, clause 12.2.1 of BS 5839-1 requires fault indication at the control and indicating equipment (which is typically sited in the entrance lobby of a building, security office or some such similar location likely to be occupied at all material times) to occur within 100 seconds if the following were to occur:

BS 5839-1

  ▶ a short-circuit or open circuit of any
    − circuits supplying fire alarm devices
    − wiring between a power supply and any equipment which it supplies that is housed in a separate enclosure
    − wiring between the control and indicating equipment and any remote control/ indicating equipment, such as repeater panels
    − wiring between control equipment and any equipment in a separate enclosure for the transmission of signals to an alarm receiving centre
  ▶ any earth fault that would be capable of preventing the fire alarm system from operating in accordance with the recommendations of BS 5839-1.

If a safety supply is derived from an IT system, continuous insulation monitoring should be provided such that an audible and visual indication is given when a first fault occurs.

7.3   Sources for safety services

560.6.1   Any equipment forming part of a source for a safety service must be installed as fixed
560.6.2   equipment in a location that is only accessible to skilled or instructed persons.

7.3.1 Generators

560.6.3   Consideration should be given at the design stage to ensure that a source of supply such as an oil-powered generator is installed such that any gases, vapours, smoke or fumes emitted cannot access parts of a building which are intended to be routinely occupied by persons. This will require the provision of suitable ventilation arrangements including, in some cases, the installation of extraction plant.

560.6.4   The use of two supplies derived from a public supply network should not be selected as safety supply services unless it can be confirmed that they will not be prone to fail at the same time. As such, this would require discussion between the designer and the distributor to ensure that the supplies were appropriately configured.

Any safety source should be capable of supporting at least all of the safety services within a building. However, a safety source may supply additionally other electrical loads within the premises where this does not detrimentally impact upon the supply required by the safety services. As such, faults occurring on a non-safety service supplied from a safety source on failure of the normal supply arrangements must not result in the disconnection of a safety service. In practice, care must be exercised when arranging the protective measures for all circuits to ensure that the protective device closest to a fault will operate, minimising disruption to and disconnection of the healthy circuits and preventing danger being caused by the loss of safety services. Consideration should also be given to arrangements for load-shedding to ensure that safety services are given priority in the event of failure of the normal supply.

560.6.5
560.6.6

314.1

Where a generator providing an alternative source for a safety supply cannot operate in parallel with the normal supply it is necessary to ensure that the relevant requirements of Section 551 are met. These include:

560.6.7

▶ The generator must have its own earthing arrangements independent of those derived from the earthed point of a TN system forming part of a public supply, such that the generator may be operated safely in the event of loss of the public supply earthing arrangement

551.4.3.2.1

▶ Precautions must be taken to prevent the generator being connected in parallel with the normal supply derived from a public supply network. This may be achieved by employing one or more of the following measures:
  – Electrical, mechanical or electromechanical interlocks
  – A system of locks with a single transferable key
  – A three-position break-before-make changeover switch
  – An automatic changeover switch with suitable interlock
  – Other suitable means which prevent parallel operation.

551.6.1

In TN-S system installations in situations where the neutral is not isolated, care must be taken to ensure that RCDs are so positioned that they do not operate inadvertently as a result of any parallel neutral-earth path. Consideration should also be given to isolation of the installation neutral from the public supply neutral in TN system installations as a means of avoiding disturbances arising from induced voltages caused by lightning being introduced into the installation.

551.6.2

Where a generator providing an alternative source for a safety supply is capable of operating in parallel with the normal supply it is necessary to ensure that the relevant requirements of Section 551 are met.

560.6.8

For example, the designer must confirm that the measures employed for protection against thermal effects and protection against overcurrent will remain effective at all times under all modes of operation. If the generator is placed on the supply side of all the protective devices for the final circuits of the installation, no special measures are required. If, however, the generator is placed on the load side of all the protective devices for a final circuit of the installation, the conductors of the final circuit in question should meet the following requirement:

551.7.1

551.7.2

$$I_z \geq I_n + I_g$$

where:

I_z is the current-carrying capacity of the final circuit conductors
I_n is the rated current of the protective device of the final circuit
I_g is the rated output current of the generating set.

All of the other requirements of Regulation 551.7.2 must also be met.

560.6.12
BS 7698-12
Any generator forming a part of an electrical source for a safety service should comply with the relevant clauses of BS 7698-12:1998 *Reciprocating internal combustion engine driven alternating current generating sets. Emergency power supply to safety devices.*

7.4 Circuits

560.7.1
Circuits supplying safety services should be independent of other circuits such that a fault, modification work or maintenance on one type of circuit does not have a detrimental effect on any other. This is generally achieved by the use of independent wiring systems for the safety services, or suitably compartmentalised wiring systems.

560.7.2
Circuits should, wherever possible, be routed to avoid their passing through locations that present a fire risk. However, where this cannot be avoided the cables must be fire-resistant. Circuits of safety services should not be routed through areas of a building that are exposed to an explosion risk.

560.7.3
433.3.3
Generally, circuits for safety services should be provided with overload protection. However, Regulation 560.7.3 refers to Regulation 433.3.3 and permits the omission of overload protection where the loss of supply under such conditions would in itself present a greater hazard. Where protection against overload is omitted, the occurrence of an overload should be indicated in some manner such as warning lights and/or supervisory buzzers.

560.7.4
Given the importance of maintaining the supply to safety services, care should be taken at the design stage to ensure that an overcurrent in one circuit has no detrimental effect such as unwanted disconnection of other circuits supplying safety services.

560.7.6
Further, where such equipment is supplied from two different circuits, they should be so arranged that the supply to the equipment is maintained if a fault were to occur on either of the circuits. A fault occurring on either circuit should not detrimentally affect the measures for protection against electric shock or the correct operation of the remaining circuit.

BS 5839-6
It should be noted that clause 15.5 a) ii) of BS 5839-6:2004 permits domestic fire detection/alarm devices forming part of a grade D system (as defined in BS 5839-6) to be supplied from a 'separately electrically protected or regularly used local lighting circuit'. Reference should also be made within this publication to paragraph 1.19 of Appendix B (for installations in England and Wales) and clause 2.11.2 of Appendix E (for installations in Scotland).

560.7.7
528.1
BS 5839-1
It is necessary for cabling other than metallic screened fire-resistant cables forming part of a safety circuit to be effectively and reliably separated from the cabling of other circuits in accordance with the general requirement given in Regulation 528.1 and any specific requirements of the relevant British Standard (such as, in the case of fire alarm systems, those given in clause 26.2.k of BS 5839-1).

Clause 26.1 of BS 5839-1:2002 (2008) states that 'the circuits of fire alarm systems need to be segregated from the cables of other circuits to minimize any potential for other circuits to cause malfunction of the fire alarm system'.

As was the case in the 16th Edition, it remains unacceptable to install any wiring other than that directly associated with the fire rescue service lift within a lift shaft or any similar flue-like shaft.

560.7.8

7.5 Information to be made available

Chapter 56 contains a number of requirements regarding information that must be made available to and remain accessible to the occupants of a building and indeed emergency services who may be required to attend the premises in the case of fire. These are summarised below.

▶ Any switchgear/controlgear associated with the supply of a safety service should be clearly identified. Any such switchgear should also be located in a plant room or similar such that it is only accessible to skilled or instructed persons

560.7.5

▶ A general schematic diagram and details of the supply arrangements of the electrical safety services should be posted in close proximity to the relevant distribution board. The diagram need not be overly complex, but its content must be maintained as and when changes to the supply are made, so that it remains an accurate reflection of the system as installed

560.7.9

▶ Copies of all relevant installation drawings should be posted at the origin of the installation such that the location of the following items can be pinpointed:
  – All electrical control equipment including distribution boards
  – Items of safety equipment with the relevant final circuit designation. The particulars and purpose of the equipment should be evident
  – Any special switching and monitoring equipment such as area switches and visual and/or acoustic warning devices relating to the safety power supply

560.7.10

▶ A list should be provided detailing all current-using equipment which is permanently connected to a safety power supply. Relevant information relating to rated currents, starting currents and starting times should be provided

560.7.11

▶ The operating instructions as appropriate for the particular installation conditions should be provided for any items of safety equipment and electrical safety services.

560.7.12

7.6 Wiring systems

Wiring systems forming part of a safety service which is required to operate under fire conditions need to comply with Regulation 560.8.1. In the case of emergency lighting installations the requirements of BS 5266-1:2005 should be met and in the case of fire detection and alarm systems in buildings the requirements of BS 5839-1:2002 (2008) apply.

560.8.1
BS 5266-1
BS 5839-1

To meet the requirements of Regulation 560.8.1, the cables forming part of a safety service which is required to operate under fire conditions need to comply with either BS EN 50362 or BS 8491 if over 20 mm overall diameter, or BS EN 50200 if of overall diameter less than or equal to 20 mm, in terms of fire resistance.

BS EN 50362
BS 8491
BS EN 50200

Additionally, in either case they should also comply with BS EN 60332-1-2 for resistance to flame.

BS EN 60332-1-2

Other cables deemed to offer the necessary degree of fire and mechanical protection may also be used, in which case reference should be made to the recommendations of the appropriate British Standard.

The following types of cable are recommended for use for the wiring forming part of critical signal paths and low voltage and/or extra-low voltage supplies to fire alarm systems (clause 26.2 of BS 5839-1 refers), or for the wiring between a central battery source and the emergency escape luminaires it supplies (clause 9.2.2a of BS 5266-1):

BS 5839-1
BS 5266-1

BS EN 60702-1
BS EN 60702-2

- ▶ Mineral insulated cable to BS EN 60702-1 installed with terminations to BS EN 60702-2

BS 7629
- ▶ Cables conforming to BS 7629

BS 7846
- ▶ Cables conforming to BS 7846.

In all cases and regardless of the type of cable installed, cables must be installed in such a way that the circuit remains electrically functional for as long as possible under fire conditions. Clause 26.2.f) of BS 5839-1:2002 (incl AMD 2 2008) specifically precludes the use of plastic cable clips, ties and trunking as the sole means of supporting cables forming part of the wiring of a fire alarm system both as a means of maintaining circuit integrity and to minimise the hazard presented by unsecured cables under fire conditions. The preclusion on the use of plastic clips, ties and the like as the sole means of cable support also appears in clause 9.2.2 of BS 5266-1.

BS 5839-1

BS 5266-1

It should be noted at this point that minimum conductor sizes are stated for cables in both emergency lighting and fire alarm systems including those used for supplies as an aid to maintaining circuit integrity as follows:

- ▶ 1 mm^2 for fire alarm systems (clause 26.2j of BS 5839-1 refers)
- ▶ 1.5 mm^2 for emergency lighting (clause 9.2.3 of BS 5266-1 refers)

560.8.2 Attention should be paid to the need to separate/segregate any cables or bus systems which might adversely affect the operation of the safety service, in accordance with the general requirements of Section 528 (of BS 7671) and any particular requirements of the appropriate British Standard. For fire alarm systems refer to clause 26n of BS 5839-1 and for emergency lighting installations see clause 9.2.5 of BS 5266-1.

Section 528

560.8.3 Care should be taken to ensure that any wiring forming part of the supply to the safety service is installed in such a way as to minimise the risk of short-circuit or earth fault, or the risk of fire or danger to persons. In some cases and with some types of cable, this may require some thought to be given to the routes taken by cables or the installation of additional mechanical protection. Again reference should be made to the appropriate British Standard, and any particular recommendations therein should be taken into consideration.

560.9
560.10

Special installations and locations 8

8.1 Introduction

Although the general requirements of BS 7671, including those relating to protection against fire, will apply to all electrical installations, these may be added to or amended by more specific requirements considered to be appropriate in the case of special installations or locations.

Section 700

BS 7671 now contains requirements for fourteen types of special installation or location and the following four contain additional requirements for protection against fire:

- ▶ Agricultural and horticultural premises
- ▶ Exhibitions, shows and stands
- ▶ Temporary electrical installations for structures, amusement devices and booths at fairgrounds, amusement parks and circuses
- ▶ Floor and ceiling heating systems.

Section 705
Section 711
Section 740

Section 753

These are considered below.

8.2 Agricultural and horticultural premises

Section 705

8.2.1 The risks

Agricultural premises frequently contain buildings that may be used for a wide range of activities which could present or contribute to a risk of fire.

Given the characteristic high level roofs/ceilings found in many storage buildings on farms, the use of projecting beam flood and spot lights is commonplace. Parts of the premises may be used for storage, refuelling, maintenance and repair of vehicles and machinery. Whilst storage of most fuels in itself is not considered too hazardous, handling or refuelling may well be. The types of maintenance and repair activities being carried out could reasonably include hot work such as welding or brazing. Other buildings may be used for the large-scale storage of fertilisers, which can under fire conditions act as oxidising agents, flammable feeds and bedding materials such as hay and straw, which in themselves can present a fire risk if incorrectly stored even without the presence of an electrical installation.

Agricultural and horticultural premises, in particular areas used for storage of edible items, are likely to attract rodents such as mice and rats which are known to gnaw on cables, causing damage to the insulation, with the risks that this presents for short-circuits or earth faults to occur. These in turn could produce arcing sufficient to present a risk of ignition.

However, to make matters worse it can be quite common for buildings in agricultural premises to be used for any number of the purposes mentioned above at the same time, presenting a combination of flammable or combustible materials, oxidising agents, raised temperatures and the presence of unshielded arcs and sparks.

131.1   It should be remembered that any livestock present on a farm must also be safeguarded from excessive temperatures likely to be caused by heating appliances or lighting.

8.2.2 Radiant heaters

705.422.6   Any electrical heating appliances which are employed in areas used for housing livestock
BS EN 60335-2-71   should meet the requirements of BS EN 60335-2-71:2003 *Household and similar electrical appliances. Safety. Particular requirements for electrical heating appliances for breeding and rearing animals.* Any heaters installed in such locations should be installed in accordance with the specified clearance from the livestock or combustible
510.2   materials stated by the manufacturer. In the absence of any such recommendation a minimum clearance of 0.5 m from livestock or combustible materials must be maintained.

8.2.3 Employment of RCDs as a measure for protection against fire

705.422.7   The 17th Edition recognises the use of RCDs having a rated residual operating current not exceeding 300 mA which disconnects all live conductors as a measure for protection against fire. This is a notable change from the 16th Edition requirements under which an RCD of 500 mA rated residual operating current could be employed for this purpose. The 17th Edition also accepts the use of 'S' type or time delay RCDs for fire protection purposes where the circuit being so protected does not supply socket-outlets.

It is highly likely that most circuits within an agricultural or horticultural installation will already be subject to RCD protection as a means of providing the protective measure of automatic disconnection of supply, in accordance with the requirements of Regulation 411.1. Where this is the case, RCDs providing automatic disconnection of supply can also be deemed to provide the required protection against fire.

8.2.4 Additional requirements for ELV circuits

705.422.8   In any location where a risk of fire is considered to exist, any conductors of a circuit supplied at extra-low voltage should be provided with mechanical protection by means of barriers or enclosures such that a level of protection against ingress of IPXXD or IP4X is achieved, or be otherwise installed in a wiring system (i.e. conduit, trunking or similar) made from insulating materials.

It should be noted that type H07RN-F heavy duty rubber industrial flexible cables meet the requirements of this regulation without the need for any additional measures being applied.

8.2.5 Rodent activity

705.513.2   Whilst Regulation 705.513.2 requires electrical equipment to be placed beyond the reach of livestock likely to be present, this requirement relates only to animals which are intended to be present and not to vermin such as rats, mice or squirrels which are likely to be present on account of the presence of shelter and a source of food. As
705.522.10   such, Regulation 705.522.10 contains a particular requirement that special attention be given to the presence of rodents.

Besides the risk of cable damage through gnawing mentioned earlier, rodents are known to nest on top of cable runs and within incorrectly sealed enclosures which again in itself may present a risk of fire.

Particular attention should be paid to the sealing arrangements applied to ducts and other cableways to prevent access for animals. Any cables considered to be so placed as to be vulnerable to the gnawing activities of rodents should be suitably protected from such actions by the use of steel trunking, conduit and accessory boxes, or possibly high impact plastic conduit, trunking etc. It is unlikely that further protection would be required for steel wire armoured or mineral insulated cables, although, given sufficient time, even these types of cable would not be beyond suffering serious damage.

8.2.6 Luminaires

Luminaires should be installed in accordance with the general requirements contained in Section 559 and Chapter 42 as appropriate (see Chapter 3 of this publication).

Section 559
Chap 42

When luminaires are being selected for use in agricultural or horticultural premises consideration must be given to:

705.559

▶ the degree of protection offered against ingress of dust, solid objects and moisture, an IP rating of IP54 being a reasonable minimum standard
▶ their suitability for installation directly attached to a normally flammable surface, such luminaires being marked ▽F (see section 3.2.4 of this publication for further information relating to changes in markings used for luminaires etc.) and of a type having a limited surface temperature marked ▽D .

8.3 Exhibitions, shows and stands

BS 7671 now considers the particular issues present in electrical installations for exhibitions, shows and stands. In the case of protection against fire, Regulation 711.422.4.2 contains the following requirements:

711.422.4.2

▶ Any lighting equipment or other equipment being used within the location having a high surface temperature under conditions of normal operation should be so placed or guarded that it does not constitute a fire hazard
▶ Any signs or similar should be sufficiently robust in terms of heat resistance, mechanical strength, electrical insulation employed and ventilation such that they can adequately withstand heat likely under normal operating conditions
▶ Where a stand contains a number of items of electrical equipment, lamp types likely to result in the production of excessive levels of heat should not be installed unless ventilation sufficient to alleviate problems regarding overheating is provided.

Regulation 711.55.1.5 requires that equipment such as heaters and luminaires which may focus, direct or concentrate heat should be fixed and protected such that the heat emitted does not constitute an ignition hazard to any materials within the location.

711.55.1.5

There is also a requirement for luminaires installed within arm's reach (2.5 m) of floor level to be securely fixed and suitably mounted and/or guarded to prevent a risk of injury including burns or ignition of materials from occurring.

711.559.5

8.4 Temporary electrical installations for structures, amusement devices and booths at fairgrounds, amusement parks and circuses

740.422.3.7 — Another section new to the 17th Edition, Section 740, contains a specific requirement that any motor which is either automatically or remotely controlled but which is not under continuous supervision should be fitted with an excess temperature cut-out that requires manual resetting.

This measure is intended to ensure that an examination of the motor and its immediate surroundings is made prior to the supply being reinstated after the cut-out has operated.

740.55.1.5 — Regulation 740.55.1.5 requires that any luminaires and floodlights be so placed that they do not constitute an ignition hazard to materials in the location as a result of the effects of focusing or concentration of the heat that they emit in normal use.

8.5 Floor and ceiling heating systems

Section 753
753.423 — The 17th Edition also sees the introduction of a new section for floor and ceiling heating systems. This section contains a specific requirement limiting the surface temperature of floors likely to be trodden on to a reasonable level of, for example, not more than 35 °C.

753.424.1.1 — The floor or ceiling heating system itself should be installed in accordance with the manufacturer's instructions including, where necessary, the use of temperature limiting devices, such that its output is limited to a maximum of 80 °C.

Further, heating units should be connected to the fixed installation from which they are supplied by cold tails (as defined in Part 2), or via terminals having an appropriate temperature rating. These measures are intended to prevent unwanted transference of heat such that damage or deterioration could occur to the insulation of the fixed wiring.

Where cold tails are employed, these should be permanently attached to the heating unit by crimped connections or other means that cannot be readily disconnected and that will not be adversely affected by the temperatures likely to be encountered in normal service.

753.424.1.2 — As it is possible that higher temperatures or indeed arcing may occur under fault conditions, attention should be paid to the provision of appropriate measures to meet the requirements of Chapter 42 (Thermal effects). Where the heating element is to be installed in close proximity to parts of the building fabric considered easily ignitable, the heating element should be placed on a non-flammable layer (such as metal sheeting), or in steel conduit, or a minimum clearance of 10 mm maintained. Other measures providing an equivalent degree of segregation are not excluded.

753.522.1.3 — Consideration should be given to the likely increase in ambient temperature where circuit wiring and control leads are installed in close proximity to heated surfaces.

GN7 — Further coverage of electrical installations in special installations or locations can be found in Guidance Note 7.

Safety service and product standards of relevance to this publication

A

BS or EN number	Title	References in this Guidance Note
BS 476-4:1970	Fire tests on building materials and structures. Non-combustibility test for materials	3.8
BS 476-12:1991	Fire tests on building materials and structures. Method of test for ignitability of products by direct flame impingement	3.2.1 3.8
BS 4086:1966	Recommendations for maximum surface temperatures of heated domestic equipment	Chap 5
BS 4884 series 1992 and 1993	Technical manuals	Introduction
BS 4940 series 1994	Technical information on construction products and services	Introduction
BS 5266-1:2005	Emergency lighting. Code of practice for the emergency lighting of premises	4.2 7.2 7.6 Appx E 2.10.3
BS 5266-7:1999	Lighting applications. Emergency lighting (also known as BS EN 1838:1999)	7.2 Appx E 2.10.3
BS 5446-1:2000	Fire detection and fire alarm devices for dwellings. Specification for smoke alarms (this standard has been superseded/withdrawn, but will remain referenced in Approved Document B until it is next revised)	Appx B Vol. 1, 1.4 Appx B Vol. 2, 1.5
BS 5446-2:2003	Fire detection and fire alarm devices for dwellings. Specification for heat alarms	Appx B Vol. 1, 1.4 Appx B Vol. 2, 1.5
BS 5839-1:2002 +A2:2008	Fire detection and fire alarm systems for buildings. Code of practice for system design, installation, commissioning and maintenance	4.2 7.2 7.4 7.6 Appx B Appx E 2.11.1 Non-Dom
BS 5839-3:1998	Fire detection and alarm systems for buildings. Specification for automatic release mechanisms for certain fire protection equipment	Appx E 2.1.14 Non-Dom Appx E 2.2.9 Dom
BS 5839-6:2004	Fire detection and fire alarm systems for buildings. Code of practice for the design, installation and maintenance of fire detection and fire alarm systems in dwellings	7.4 Appx B Appx E 2.11.2 Dom Appx E 2.11.3 Dom

continues

BS or EN number	Title	References in this Guidance Note
BS 5839-8:2008	Fire detection and fire alarm systems for buildings. Code of practice for the design, installation, commissioning and maintenance of voice alarm systems	Appx B Vol. 2, 1.32
BS 6351-2:1983	Electric surface heating. Guide to the design of electric surface heating systems	3.2.1
BS 6351-3:1983	Electric surface heating. Code of practice for the installation, testing and maintenance of electric surface heating systems	3.2.1
BS 6724:1997 +A3:2008	Electric cables. Thermosetting insulated, armoured cables for voltages of 600/1000 V and 1900/3300 V, having low emission of smoke and corrosive gases when affected by fire	Chap 2
BS 7211:1998	Electric cables. Thermosetting insulated, non-armoured cables for voltages up to and including 450/750 V, for electric power, lighting and internal wiring, and having low emission of smoke and corrosive gases when affected by fire	Chap 2
BS 7273-1:2006	Code of practice for the operation of fire protection measures. Electrical actuation of gaseous total flooding extinguishing systems	Appx B Vol. 2
BS 7273-2:1992	Code of practice for the operation of fire protection measures. Mechanical actuation of gaseous total flooding and local application extinguishing systems	Appx B Vol. 2
BS 7273-4:2007	Code of practice for the operation of fire protection measures. Actuation of release mechanisms for doors	Appx B Vol. 2
BS 7273-5:2008	Code of practice for the operation of fire protection measures. Electrical actuation of watermist systems (except pre-action systems)	Appx B Vol. 2
BS 7540 series:2005	Electric cables. Guide to use for cables with a rated voltage not exceeding 450/750 V	Chap 2
BS 7629-1:2008	Electric cables. Specification for 300/500 V fire resistant screened cables having low emission of smoke and corrosive gases when affected by fire. Multicore and multipair cables	Chap 2 7.6
BS 7671:2008	Requirements for Electrical Installations. IEE Wiring Regulations. Seventeenth Edition	Numerous
BS 7698-12:1998	Reciprocating internal combustion engine driven alternating current generating sets. Emergency power supply to safety devices (also known as ISO 8528-12:1997)	7.3
BS 7846:2000	Electric cables. 600/1000 V armoured fire-resistant cables having thermosetting insulation and low emission of smoke and corrosive gases when affected by fire	Chap 2 7.6
BS 8491:2008	Method for assessment of fire integrity of large diameter power cables for use as components for smoke and heat control systems and certain other active fire safety systems	4.6 7.6
BS EN 54-2:1998	Fire detection and fire alarm systems. Control and indicating equipment	Appx B Vol. 1, 1.6; 1.7 Appx B Vol. 2, 1.4
BS EN 54-4:1998	Fire detection and fire alarm systems. Power supply equipment	Appx B Vol. 1, 1.6; 1.7 Appx B Vol. 2, 1.4
BS EN 54-11:2001	Fire detection and fire alarm systems. Manual call points	Appx B Vol. 2, 1.31 Appx E 2.11.1
BS EN 50085 series	Cable trunking systems and cable ducting systems for electrical installations	4.2 4.3.4 6.1

BS or EN number	Title	References in this Guidance Note
BS EN 50085-1:2005	Cable trunking systems and cable ducting systems for electrical installations. General requirements	4.4.3
BS EN 50086 series	Specification for conduit systems for cable management	6.1
BS EN 50200:2006	Method of test for resistance to fire of unprotected small cables for use in emergency circuits	4.6 7.6
BS EN 50266 series: 2001	Common test methods for cables under fire conditions. Test for vertical flame spread of vertically mounted bunched wires or cables	4.2 4.3.4 6.1
BS EN 50362:2003	Method of test for resistance to fire of larger unprotected power and control cables for use in emergency circuits	7.6
BS EN 60079-14:2008	Explosive atmospheres. Electrical installations design, selection and erection	4.3 7.1
BS EN 60332-1-2:2004	Tests on electric and optical fibre cables under fire conditions. Test for vertical flame propagation for a single insulated wire or cable. Procedure for 1 kW pre-mixed flame	4.3.4 4.4.3 6.1 7.6
BS EN 60439-2:2000	Low voltage switchgear and controlgear assemblies. Particular requirements for busbar trunking systems (busways). (also known as IEC 60439-2:2000)	4.3.4 6.1
BS EN 60529:1992 (2004)	Specification for degrees of protection provided by enclosures (IP code)	4.3.3
BS EN 60598-1:2004	Luminaires. General requirements and tests	3.2.4 4.4.2
BS EN 60598-1:2008	Luminaires. General requirements and tests	3.2.4
BS EN 60598-2-23: 1997	Luminaires. Particular requirements. Extra low voltage lighting systems for filament lamps (also known as IEC 60598-2-23:1996)	3.2.3
BS EN 60598-2-24: 1999	Luminaires. Particular requirements. Luminaires with limited surface temperatures	4.3.1
BS EN 60702-1:2002	Mineral insulated cables and their terminations with a rated voltage not exceeding 750 V. Cables (also known as IEC 60702-1:2002)	Chap 2 4.3.4 4.6 7.6
BS EN 60702-2:2002	Mineral insulated cables and their terminations with a rated voltage not exceeding 750 V. Terminations (also known as IEC 60702-2:2002)	7.6
BS EN 61034-2:2005	Measurement of smoke density of cables burning under defined conditions. Test procedure and requirements	4.2
BS EN 61184:1997	Bayonet lampholders (also known as IEC 61184:1997)	3.4
BS EN 61386-1:2008	Conduit systems for cable management. General requirements (BS EN 61386-1:2004 also remains current)	4.2 4.3.4 4.4.3
BS EN 61534 series	Powertrack systems	4.3.4 6.1
BS EN 61537:2007	Cable management. Cable tray systems and cable ladder systems	4.3.4 6.1

Approved Document B1: Means of warning and escape   **B**

For Scotland see Appendix E of this publication.

For Northern Ireland see the relevant deemed-to-satisfy provisions of the publications referred to in Table E to Schedule 5 of the Building Regulations (Northern Ireland) 2000 as amended. The publications include Technical Booklet E:2005 and other fire safety documents. The legislation and technical booklet can be found on the Department of Finance and Personnel's building regulations website: www.buildingregulationsni.gov.uk

Note: Approved Document B (Fire safety) has been subdivided into two volumes. Volume 1 contains guidance applicable to dwellinghouses while Volume 2 contains guidance applicable to buildings other than dwellinghouses. As such, the guidance below is similarly divided along these lines.

In the text below, Approved Document B is abbreviated to ADB.

1.1 (ADB Vols. 1 and 2)
Section 1 of both volumes prescribes measures to be taken in buildings with respect to fire detection and alarm systems in order to give early warning in the event of fire.

Volume 1: Dwellinghouses

Note: Volume 1 is relevant to houses and the individual residential units of sheltered housing. For guidance applicable to the common parts of a sheltered housing development, flats, student accommodation and similar buildings, reference should be made to Volume 2.

1.2 (ADB Vol. 1)
It is recognised that the installation of either smoke alarms or automatic fire detection and alarm systems can significantly increase the level of safety by their raising the alarm automatically in the event of a fire breaking out. In most cases the guidance given in Volume 1 will be appropriate; however, a higher standard of protection might be necessary where occupants are at particular risk. As such, the approved document gives minimum recommendations for typical domestic situations.

1.3 (ADB Vol. 1)
All new dwellinghouses should be provided with a fire detection and fire alarm system. This system should meet, as a minimum, the recommendations of BS 5839-6:2004 for a Grade D, LD3 standard.

BS 5839-6

BS 5839-6:2004 *Code of practice for the design, installation and maintenance of fire detection and fire alarm systems in dwellings* defines a Grade D system as 'A system of one or more mains-powered smoke alarms, each with an integral standby supply (the system may, in addition, incorporate one or more mains-powered heat alarms, each with an integral standby supply)' and Category LD3 as 'A system incorporating detectors in all circulation spaces that form part of the escape routes from the dwelling.'

1.4 (ADB Vol. 1)

Any smoke and heat alarms used should be mains-operated and have a standby supply from a battery or capacitor.

BS 5446-1   Smoke alarms should conform to BS 5446-1:2000 *Fire detection and fire alarm devices – Specification for smoke alarms.*

BS 5446-2   Heat alarms should conform to BS 5446-2:2003 *Fire detection and fire alarm devices – Specification for heat alarms.*

1.5 (ADB Vol. 1)

A dwellinghouse is considered to be large if it has more than one storey and any such storey exceeds 200 m².

1.6 (ADB Vol. 1)

BS 5839-6   A large dwellinghouse having two storeys (not including any basement storey) should be fitted with a fire detection and fire alarm system which meets, as a minimum, the recommendations of BS 5839-6:2004 for a Grade B, Category LD3 system.

BS EN 54-2
BS EN 54-4   BS 5839-6:2004 defines a Grade B system as 'A fire detection and fire alarm system comprising fire detectors (other than smoke alarms and heat alarms), fire alarm sounders, and control and indicating equipment that either conforms to BS EN 54-2 (and power supply complying with BS EN 54-4) or to Annex C of this part (6) of BS 5839.'

A Category LD3 system is defined as 'A system incorporating detectors in all circulation spaces that form part of the escape routes from the dwelling.'

1.7 (ADB Vol. 1)

BS 5839-6   A large dwellinghouse having three or more storeys (not including any basement storey) should be fitted with a fire detection and fire alarm system which meets, as a minimum, the recommendations of BS 5839-6:2004 for a Grade A, Category LD2 system.

BS 5839-6:2004 defines a Grade A system as 'A fire detection and fire alarm system, which incorporates control and indicating equipment conforming to BS EN 54-2, and power supply equipment conforming to BS EN 54-4…' In the most part, such a system should be designed and installed in accordance with the relevant parts of BS 5839-1. However, those aspects of the design/installation relating to audible alarm devices and audibility; fire alarm warnings for persons who are deaf or hard of hearing; manual call points; capacity of standby batteries; and radio linked systems, should be in accordance with the relevant parts of BS 5839-6. For more precise information on this matter, refer to clause 7.1 of BS 5839-6.

A Category LD2 system is defined as 'A system incorporating detectors in all circulation spaces that form part of the escape routes from the dwelling, and in all rooms or areas that present a high fire risk to occupants…'

1.8 (ADB Vol. 1)

Where alterations or additions are made to a domestic dwelling such that new habitable rooms are provided above ground floor level, or at ground floor level and there is no final exit from these newly provided rooms, it will be necessary to install a fire detection and alarm system. Smoke alarms will be required in the circulation spaces in accordance with paragraphs 1.10 to 1.18, guidance on which is given below.

The intention is that occupants of these new rooms will receive warning of a fire occurring in areas forming part of their escape route from the premises.

1.9 (ADB Vol. 1)

The fire detection and alarm systems in the individual dwellings in a sheltered housing scheme under the control of a warden or supervisor should be connected to a central monitoring point or alarm receiving centre. It should be possible for the person watching over the monitoring for the sheltered housing complex to identify in which individual dwelling(s) a fire alarm has been raised.

It should be remembered that communal areas of a sheltered housing scheme should meet the recommendations of Volume 2 of Approved Document B, as should the sheltered accommodation of institutional/residential premises such as nursing accommodation, police section houses and the like.

1.10 (ADB Vol. 1)

In general the design and installation of the fire detection and alarm systems in dwellinghouses should be in accordance with the recommendations given in BS 5839-6:2004. However, the following points are emphasised in ADB Volume 1.

BS 5839-6

1.11 (ADB Vol. 1)

In order that an alarm is raised in the early stages of a fire, smoke alarms should be installed in circulation spaces such as corridors, hallways or landings between bedrooms, and those places where fires are most likely to start (that is, kitchens and living rooms).

1.12 (ADB Vol. 1)

At least one smoke detector should be installed on every storey.

1.13 (ADB Vol. 1)

An open plan situation where the kitchen area is not separated from stairways or circulation spaces by a door requires a compatible heat detector or heat alarm (in terms of the system to which it is to be connected) to be placed in the kitchen area, which should be interconnected to the smoke detector(s) installed in the circulation areas.

1.14 (ADB Vol. 1)

Where the alarm system consists of more than one detector/sounder unit, they should be linked so that detection of smoke or heat by one unit raises the alarm in all of them simultaneously. Any manufacturers' recommendations regarding the maximum number of units that can be linked should be observed.

1.15 (ADB Vol. 1)

The following should be observed when siting detectors on typical flat ceilings:

▶ In a circulation area there should be a smoke detector within 7.5 m of the door of every habitable room

▶ Detectors designed to be ceiling mounted should be mounted at least 300 mm from walls and light fittings (unless there is test evidence to demonstrate that such a proximity to light fittings will not adversely affect the functioning of the detector)

▶ The sensor in ceiling mounted smoke detector/sounder units should be between 25 mm and 600 mm below the ceiling

▶ The sensor in ceiling mounted heat detector/sounder units should be between 25 mm and 150 mm below the ceiling

Detectors designed for wall mounting may also be used if installed above the level of doorways opening into the circulation space and fixed in accordance with manufacturers' instructions.

1.16 (ADB Vol. 1)

Smoke alarm/detector units should be sited in easily accessible positions so that routine maintenance activities such as testing and cleaning can be carried out safely. They should not be placed over stairs or any openings between floors.

The guidance given in paragraphs 1.3 to 1.6 in ADB Vol. 1 and above is in line with the guidance and recommendations given in BS 5839-6 clauses 11.1.1 and 11.2.

1.17 (ADB Vol. 1)

In order to minimise the likelihood of false alarms, smoke detectors/alarms should **not** be installed:

▶ next to or directly over heaters or air-conditioning units/outlets

▶ in bathrooms

▶ in showers

▶ in kitchens/cooking areas

▶ in garages

▶ in locations where steam, condensation or fumes are likely to be present.

1.18 (ADB Vol. 1)

Similarly, smoke detectors/alarms should **not** be installed in locations that:

▶ get very hot, such as boiler rooms and laundry areas,

▶ can be very cold, such as unheated porches.

Fixing smoke detectors/alarms to surfaces which are likely to be considerably warmer or colder than their immediate surroundings should be avoided as this would present the possibility of air currents developing which may prevent or delay smoke from entering the detector unit.

The guidance given in paragraphs 1.17 and 1.18 in ADB Vol. 1 and above is in line with the guidance and recommendations given in BS 5839-6 clauses 12.1 and 12.2.

1.19 (ADB Vol. 1)

The power supply to the smoke detector/alarm system should come from either an independent circuit originating from the main distribution board, or a regularly used local lighting circuit. In the case of radio-linked units, refer to paragraph 1.21 in ADB Vol. 1 and below.

Where a lighting circuit is used as the source of supply, a means to isolate the supply to the smoke detector/alarm system and not the lighting should be provided.

The logic behind permitting the supply to be taken from a local lighting circuit is that users of the property will become aware within a very short period of time if the supply is lost for any reason.

1.20 (ADB Vol. 1)

Any electrical installation work in domestic dwellinghouses should comply with Approved Document P (Electrical safety).

1.21 (ADB Vol. 1)

In general, any cable that can be used for the wiring of domestic premises may be used for the power supply for and interconnection between smoke detector/alarm units. However, in large dwellinghouses (as defined above in 1.5) BS 5839-6:2004 specifies the use of fire-resisting cables for Grade A and B systems.

Any cable used for the interconnection of smoke detector/alarm units should be readily distinguishable from cables forming part of the general low voltage electrical installation within the premises, typically by colour coding.

The interconnection between mains-powered smoke detector/alarm units may be via radio-link so long as this does not reduce the standby capacity of the units to below 72 hours' duration.

It is acceptable for units interlinked via radio-link to be supplied from a number of separate circuits.

1.22 (ADB Vol. 1)

Other options for power supplies are permitted by BS 5839-1 and BS 5839-6 and reference should be made to these standards where it is intended to use measures other than those described in 1.19 and 1.21 above.

1.23 (ADB Vol. 1)

Fire detection and alarm systems should be properly designed, installed and maintained.

Upon completion of the installation work, installation and completion certificates based on the model forms in Annexes E and F as appropriate for the grade of system installed should be provided.

It is also necessary for the installation of cables and wiring used for the power supply to, and interconnection of, smoke detector/alarm units to be inspected and tested in accordance with the requirements of BS 7671:2008, and for appropriate certification based on the model forms contained in Appendix 6 of that standard to be provided.

1.24 (ADB Vol. 1)

Any and all relevant information relating to the use and maintenance of the alarm system and its component parts should be passed on to the occupants and users of the protected premises.

For further guidance on the information required to be passed on to occupants and users, reference should be made to the following as appropriate for the grade of system installed:

- ▶ BS 5839-1:2008, section 40 *Documentation*
- ▶ BS 5839-6:2004, section 24 *User instructions*

Volume 2: Buildings other than dwellinghouses

Note: Volume 2 is relevant to buildings other than dwellinghouses. For guidance applicable to dwellinghouses and the individual dwellings forming part of controlled sheltered housing, reference should be made to Volume 1.

1.2 (ADB Vol. 2)
Paragraphs 1.3, 1.4 and 1.5 in ADB Vol. 2 and below relate to fire alarm and fire detection systems in flats.

1.3 (ADB Vol. 2)
It is recognised that the installation of an automatic fire detection and alarm system can significantly increase the level of safety in such premises by raising the alarm automatically in the event of a fire breaking out. The provisions set out in 1.4 and 1.5 may be considered to be minimum requirements. In the case of flats used to house persons considered to be at an increased risk in the event of a fire, a higher standard of protection may need to be provided.

1.4 (ADB Vol. 2)
BS 5839-6   A fire detection and alarm system should be installed in all new flats which meets, as a minimum, the recommendations of BS 5839-6:2004 for a Grade D, Category LD3. BS 5839-6:2004 defines a Grade D system as 'A system of one or more mains-powered smoke alarms, each with an integral standby supply.'

A Category LD3 system is defined as 'A system incorporating detectors in all circulation spaces that form part of the escape routes from the dwelling.'

1.5 (ADB Vol. 2)
Any smoke and heat alarms used should be mains-operated and have a standby supply from a battery or capacitor.

BS 5446-1   Smoke alarms should conform to BS 5446-1:2000 *Fire detection and fire alarm devices – Specification for smoke alarms.*

BS 5446-2   Heat alarms should conform to BS 5446-2:2003 *Fire detection and fire alarm devices – Specification for heat alarms.*

1.6 (ADB Vol. 2)
Where alterations or additions are made such that new habitable rooms are provided above ground floor level, or at ground floor level and there is no final exit from these newly provided rooms, it will be necessary to install a fire detection and alarm system. Smoke alarms will be required in the circulation spaces in accordance with paragraphs 1.10 to 1.18, guidance on which is given below.

The intention is that occupants of these new rooms will receive warning of a fire occurring in areas forming part of their escape route from the premises.

1.7 (ADB Vol. 2)
The fire detection and alarm systems in the individual dwellings in a sheltered housing scheme under the control of a warden or supervisor should be connected to a central monitoring point or alarm receiving centre. It should be possible for the person watching over the monitoring for the sheltered housing complex to identify in which individual dwelling(s) a fire alarm has been raised.

It should be remembered that communal areas of a sheltered housing scheme should meet the recommendations of Volume 2 of Approved Document B, as should the sheltered accommodation of institutional/residential premises such as nursing accommodation, police section houses and the like. The type of system to be installed in such premises should be considered on a case-by-case basis. The guidance given in paragraphs 1.24 to 1.38 in ADB Vol. 2 and described below should be taken into consideration.

1.8 (ADB Vol. 2)

In situations where up to six students share a self-contained flat having its own entrance door and where the flat has been constructed following the compartmentation principles for flats given in Section 7 of Approved Document B3: *Internal fire spread (structure)*, each flat may be provided with a separate automatic fire detection and alarm system.

In situations such as halls of residence where a general evacuation will be required, a fire detection/alarm system should be installed that follows the guidance relating to buildings other than flats given in paragraphs 1.24 to 1.38 in ADB Vol. 2 and described below.

1.9 (ADB Vol. 2)

In general the design and installation of the fire detection and alarm systems in flats should be in accordance with the recommendations given in BS 5839-6:2004. However, the following points are emphasised in ADB Volume 2.

1.10 (ADB Vol. 2)

In order that an alarm is raised in the early stages of a fire, smoke alarms should be installed in circulation spaces such as corridors, hallways or landings between bedrooms, and those places where fires are most likely to start (that is, kitchens and living rooms).

1.11 (ADB Vol. 2)

At least one smoke detector should be installed on every storey.

1.12 (ADB Vol. 2)

An open plan situation where the kitchen area is not separated from stairways or circulation spaces by a door requires a compatible heat detector or heat alarm (in terms of the system to which it is to be connected) to be placed in the kitchen area, which should be interconnected to the smoke detector(s) installed in the circulation areas.

1.13 (ADB Vol. 2)

Where the alarm system consists of more than one detector/sounder unit, they should be linked so that detection of smoke or heat by one unit raises the alarm in all of them simultaneously. Any manufacturers' recommendations regarding the maximum number of units that can be linked should be observed.

1.14 (ADB Vol. 2)

The following should be observed when siting detectors on typical flat ceilings:

▶ In a circulation area there should be a smoke detector within 7.5 m of the door of every habitable room

▶ Detectors designed to be ceiling mounted should be mounted at least 300 mm from walls and light fittings (unless there is test evidence to demonstrate that such a proximity to light fittings will not adversely affect the functioning of the detector)

▶ The sensor in ceiling mounted smoke detector/sounder units should be between 25 mm and 600 mm below the ceiling

▶ The sensor in ceiling mounted heat detector/sounder units should be between 25 mm and 150 mm below the ceiling.

Detectors designed for wall mounting may also be used if installed above the level of doorways opening into the circulation space and fixed in accordance with manufacturers' instructions.

1.15 (ADB Vol. 2)
Smoke alarm/detector units should be sited in easily accessible positions so that routine maintenance activities such as testing and cleaning can be carried out safely. They should not be placed over stairs or any openings between floors.

1.16 (ADB Vol. 2)
In order to minimise the likelihood of false alarms, smoke detectors/alarms should **not** be installed:

▶ next to or directly over heaters or air-conditioning units/outlets
▶ in bathrooms
▶ in showers
▶ in kitchens/cooking areas
▶ in garages
▶ in locations where steam, condensation or fumes are likely to be present.

1.17 (ADB Vol. 2)
Similarly, smoke detectors/alarms should **not** be installed in locations that:

▶ get very hot, such as boiler rooms and laundry areas,
▶ can be very cold, such as unheated porches.

Fixing smoke detectors/alarms to surfaces which are likely to be considerably warmer or colder than their immediate surroundings should be avoided as this would present the possibility of air currents developing which may prevent or delay smoke from entering the detector unit.

The guidance given in paragraphs 1.16 and 1.17 in ADB Vol. 1 and above is in line with the guidance and recommendations given in BS 5839-6 clauses 12.1 and 12.2.

1.18 (ADB Vol. 2)
Any and all relevant information relating to the use and maintenance of the alarm system and its component parts should be passed on to the occupants and users of the protected premises.

For further guidance on the information required to be passed on to occupants and users, reference should be made to the following as appropriate for the grade of system installed:

▶ BS 5839-1:2008, section 40 *Documentation*
▶ BS 5839-6:2004, section 24 *User instructions*

1.19 (ADB Vol. 2)
The power supply to the smoke detector/alarm system should come from either an independent circuit originating from the main distribution board, or a regularly used local lighting circuit. In the case of radio-linked units, refer to paragraph 1.21 in ADB Vol. 1 and below.

Where a lighting circuit is used as the source of supply, a means to isolate the supply to the smoke detector/alarm system and not the lighting should be provided.

The logic behind permitting the supply to be taken from a local lighting circuit is that users of the property will become aware within a very short period of time if the supply is lost for any reason.

1.20 (ADB Vol. 2)

The electrical installation should comply with Approved Document P (Electrical safety).

1.21 (ADB Vol. 2)

In general, any cable suitable for the wiring of domestic premises may be used for the power supply for and interconnection between smoke detector/alarm units.

Any cable used for the interconnection of smoke detector/alarm units should be readily distinguishable from cables forming part of the general low voltage electrical installation within the premises, typically by colour coding.

The interconnection between mains-powered smoke detector/alarm units may be via radio-link so long as this does not reduce the standby capacity of the units to below 72 hours' duration.

It is acceptable for units interlinked via radio-link to be supplied from a number of separate circuits.

1.22 (ADB Vol. 2)

Other options for power supplies are permitted by BS 5839-1 and BS 5839-6 and reference should be made to these standards where it is intended to use measures other than those described in 1.19 and 1.21 above.

1.23 (ADB Vol. 2)

Fire detection and alarm systems should be properly designed, installed and maintained.

Upon completion of the installation work, installation and completion certificates based on the model forms in Annexes E and F as appropriate for the grade of system installed should be provided.

It is also necessary for the installation of cables and wiring used for the power supply to, and interconnection of, smoke detector/alarm units to be inspected and tested in accordance with the requirements of BS 7671:2008 and for appropriate certification based on the model forms contained in Appendix 6 of that standard to be provided.

1.24 (ADB Vol. 2)

When deciding which type of fire detection/alarm system needs to be installed, consideration should be given to pertinent factors such as type of occupancy and the escape strategy that is to be employed (simultaneous, phased, or progressive horizontal evacuation).

1.25 (ADB Vol. 2)

The threat posed by fire is considered to be much greater in residential accommodation where persons are expected to sleep than in premises where the occupants are expected to be awake and alert.

Where the means of escape is based on simultaneous evacuation, operation of a manual call point ('break-glass' unit) or a detector should result in an almost instantaneous warning being raised by the sounders and other devices as appropriate.

Where the means of escape is based on phased evacuation, a staged alarm response is appropriate. Typical alarm stages might be 'alert' and 'evacuate' for example.

1.26 (ADB Vol. 2)

The type and category of fire detection/alarm system required in any given premises should be decided upon on a case-by-case basis.

General guidance on which type and category of system might be appropriate can be found in Table A1 of BS 5839-1:2002.

1.27 (ADB Vol. 2)

Arrangements should be put in place for detecting fire in all buildings. However, this does not necessarily mean the installation of an automatic fire detection/alarm system.

1.28 (ADB Vol. 2)

It is not a requirement to have an automatic fire detection/alarm system in a small non-domestic building or premises. If, as a result of carrying out a risk assessment, it can be confirmed that a suitable audible fire alarm warning could be raised by, for example, shouting, then this may be considered to be adequate. Other alternatives for raising the alarm that may prove suitable in small non-domestic locations include hand-held bells and other manually operated sounders, or an alarm system consisting of manual call points and warning devices only (that is, containing no automatic detectors) classed as a Category M system in BS 5839-1. See also item 1.30.

1.29 (ADB Vol. 2)

In larger premises not covered by 1.28 in ADB Vol. 2 and above, a building should be provided with a suitable electrically operated fire detection/alarm system containing a sufficient number of both detectors and manual call points as required to give adequate coverage of the premises, and sufficient sounders and other warning devices to provide an adequate notification of the alarm situation throughout the premises.

1.30 (ADB Vol. 2)

Any electrically operated fire detection/alarm system that is installed in non-domestic premises covered by ADB Vol. 2 should comply with all relevant parts of the current edition of BS 5839-1.

BS 5839-1 recognises three categories of system:

i Category L – offering life protection (for human occupants of premises)
ii Category P – offering property protection for the premises
iii Category M – manually operated fire alarm systems

For further information on Category M classification refer to clause 5.1.2 of BS 5939-1.

Categories L and P are further subdivided as follows:

▶ *Category L1*
A life protection system giving coverage throughout all relevant parts of the building.

▶ *Category L2*
A life protection system giving coverage only to defined parts of the protected building. Note that a Category L2 system should include the coverage required of a Category L3 system as described below.

▶ *Category L3*
A life protection system designed to give a warning of a fire sufficiently early to enable all occupants of a protected building, other than possibly those persons in the room where the fire originally occurred, to escape safely, before escape routes

become impassable through the presence of smoke, flames, toxic gases and other products of combustion.

▶ *Category L4*
A life protection system giving coverage of those parts of the escape routes comprising circulation areas and circulation spaces including corridors and stairways. The purpose of such a system is to enhance the safety of occupants by giving them early warning of the presence of smoke within escape routes through which they will need to travel to reach a place of safety.

▶ *Category L5*
A life protection system in which the protected area(s) and/or the location of the detectors is designed to meet a specific fire safety objective somewhat different to that of the Category L1, L2, L3 or L4 systems described above.

For further information on Category L classifications refer to clause 5.1.3 of BS 5939-1.

▶ *Category P1*
A property protection system installed throughout the protected building.

▶ *Category P2*
A property protection system installed only in defined parts of the building. The purpose of such a system being to allow a fire alarm to be raised as soon as possible in areas which present a high fire hazard, or in which the risk to property or business continuity is high from a fire.

For further information on Category P classifications refer to clause 5.1.4 of BS 5939-1.

1.31 (ADB Vol. 2)

Any manual call points used as part of a fire detection/alarm system should be of Type A (direct) operation as given in BS EN 54-11. A Type A call point is defined as 'a manual call point in which the change to the alarm condition is automatic (i.e. without the need for further manual action) when the frangible element is broken or displaced'. The manual call points should be installed in accordance with the relevant clauses of BS 5839-1.

BS EN 54-11

Type B operation manual call points as given in BS EN 54-11, having indirect operation where, for example, it is necessary to break the glass and then press a button to raise the alarm, may only used with the approval of the relevant local Building Control body.

1.32 (ADB Vol. 2)

In situations where it is believed that persons may not respond quickly to a fire warning, or where persons are unfamiliar with the fire warning arrangements, consideration may be given to the use of a voice alarm system. Such a system may form part of a general public address system within the building and could give both an audible signal and verbal instructions for occupants to follow in the event of a fire.

Any such warning signal should be recognisably different and distinct from any other signals in general use. Any accompanying verbal instructions should be clear.

Any installed voice alarm system should comply with the relevant clauses of BS 5839-8:2008 *Fire detection and fire alarm systems for buildings. Code of practice for the design, installation, commissioning and maintenance of voice alarm systems.*

BS 5839-8

1.33 (ADB Vol. 2)

It may not be desirable to raise a general alarm in large premises such as shopping centres and places of assembly as a result of the large number of members of the general public that are likely to be present at any given time. In such instances where a general alarm is not raised, it is important that sufficient numbers of appropriately trained staff are in place to initiate any pre-planned procedures for safe evacuation. Actuation of the fire alarm system will alert staff by means of discreet sounders, personal messages or paging or similar means.

Where full evacuation of such premises is necessary, notification can be given by sounders or a message broadcast over the public address system. The installed fire alarm system should in all other respects comply with BS 5839-1.

1.34 (ADB Vol. 2)

In situations where it is reasonably likely that one or more persons having impaired hearing are likely to be present in relative isolation such as in a hotel room whilst staying as a guest, or whilst using lavatories or in other sanitary accommodation and where no other method of alerting them is available, the audible fire alarm signal should be supplemented by a visual indication.

In buildings such as schools, colleges and offices where the persons present are under some control, the use of a vibrating pager system or similar might be appropriate.

Reference should be made to BS 5839-1, clause 18, 'Fire alarm warnings for people with impaired hearing'.

1.35 (ADB Vol. 2)

An automatic fire detection/alarm system in accordance with BS 5839-1 should be provided in all institutional and residential premises where it is expected that persons will sleep. This would include, for example, lodging blocks for students, nurses, police and the like.

1.36 (ADB Vol. 2)

Although automatic fire detection systems are not generally required in non-residential premises/buildings, there may be specific circumstances making the installation of such a BS 5839-1 type system necessary such as:

▶ To compensate for one or more departures from the guidance on fire safety given elsewhere in ADB.

▶ To act as part of the operating system for a greater fire protection system incorporating, for example, pressure differential systems or automatic door release mechanisms.

▶ Where there exists the possibility of a fire breaking out in a part of the premises which could have an adverse effect upon the means of escape from those premises. This could include locations not frequently visited in normal use such as storage areas, and parts of a building that have been vacated and are not currently in regular use.

1.37 (ADB Vol. 2)

Any installed fire detection/alarm system must be properly designed, installed and maintained. On completion, or as soon as possible after completion, of the design, installation and commissioning of the system, the appropriate certification should be completed by the person responsible for that aspect of the fire alarm system.

1.38 (ADB Vol. 2)

Fire detection/alarm systems can form a part of a larger fire protection regime within a building and may be used to initiate the operation, or change of state of other constituent parts of such a greater system such as smoke control systems; fire extinguishing systems; door release mechanisms for fire doors; automatic unlocking arrangements of exit doors, and the like.

As such, any interfaces between the fire detection/alarm system and other systems required to achieve compliance with Building Regulations must be designed to give a high degree of reliability of operation.

Particular care must be exercised if the interface is made via another system such as for example an access control system.

Where any part of BS 7273 *Code of practice for the operation of fire protection measures* applies to the actuation of other systems such as gaseous total flooding systems (Parts 1 and 2), release mechanisms for doors (Part 4) or non pre-action water mist systems (Part 5), the recommendations of the relevant part of that standard should be followed.

Guidance Note 4: Protection Against Fire

Approved Document B2: Internal fire spread (linings) C

For Scotland see Appendix E of this document.

For Northern Ireland see the relevant deemed-to-satisfy provisions of the publications of Technical Booklet E:2005 as referred to Schedule 5 of the Building Regulations (Northern Ireland) 2000 as amended. The legislation and technical booklet can be found on the Department of Finance and Personnel's building regulations website: www.buildingregulationsni.gov.uk

Note: Approved Document B (Fire safety) has been subdivided into two volumes. Volume 1 contains guidance applicable to dwellinghouses while Volume 2 contains guidance applicable to buildings other than dwellinghouses. As such, the guidance below is similarly divided along these lines.

In the text below, Approved Document B is abbreviated to ADB.

Volume 1: Dwellinghouses

Note: Volume 1 is relevant to houses and the individual residential units of sheltered housing. For guidance applicable to the common parts of a sheltered housing development, flats, student accommodation and other buildings, reference should be made to Volume 2.

Volume 2: Buildings other than dwellinghouses

Note: Volume 2 is relevant to buildings other than dwellinghouses. For guidance applicable to dwellinghouses and the individual dwellings forming part of controlled sheltered housing, reference should be made to Volume 1.

The provisions described in paragraphs 3.11, 3.12 and 3.13 in Vol. 1, and paragraphs 6.13, 6.14 and 6.15 in Vol. 2 apply to lighting diffusers which form part of a ceiling. They do not relate to diffusers of light fittings which are attached to the soffit of, or suspended beneath, a ceiling. See Figure C.1.

3.11 Vol 1
6.13 Vol 2

▼ **Figure C.1**
Lighting diffuser in
relation to ceiling

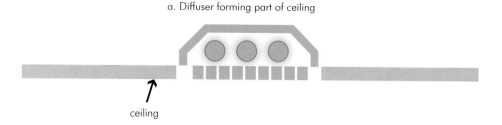

a. Diffuser forming part of ceiling

ceiling

b. Diffuser in fitting attached below and not forming part of ceiling

ceiling

3.12 Vol 1
6.14 Vol 2

Thermoplastic lighting diffusers should not be used in fire-protecting or fire-resisting ceilings, unless they have been satisfactorily tested as part of a ceiling system that can be used to provide the appropriate fire protection.

3.13 Vol 1
6.15 Vol 2

Subject to 3.11 and 3.12 in ADB Vol. 1, and 6.12 and 6.13 in ADB Vol. 2 and above, ceilings of rooms and circulation spaces (but **not** protected stairways) may contain thermoplastic lighting diffusers if the following provisions are followed:

▶ The exposed wall and ceiling surfaces above the suspended ceiling (other than the upper surfaces of the thermoplastic panels themselves) should comply with the general provisions of paragraph 3.1 of ADB Vol. 1 and paragraph 6.1 of ADB Vol. 2 (Classification of linings)
▶ For diffusers of classification TP(a) (rigid), no restrictions on their use apply
▶ For diffusers of classification TP(b), their use should be restricted in extent as given in Figure C.2 and Table C.1.

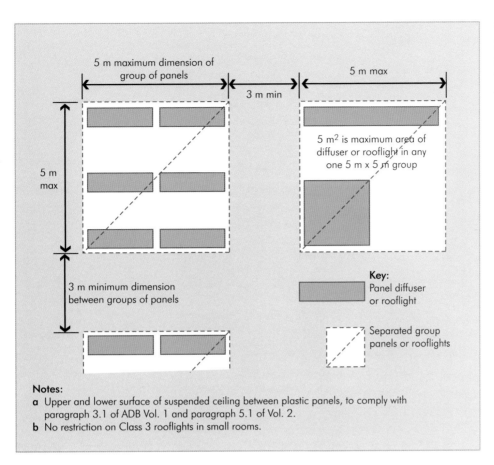

▼ Figure C.2
Layout restrictions on Class 3 plastic rooflights, TP(b) rooflights and TP(b) lighting diffusers

Notes:
a Upper and lower surface of suspended ceiling between plastic panels, to comply with paragraph 3.1 of ADB Vol. 1 and paragraph 5.1 of Vol. 2.
b No restriction on Class 3 rooflights in small rooms.

▼ Table C.1 Classification of diffusers

Minimum classification of lower surface	Use of space below the diffusers or rooflight	Maximum area of each diffuser panel or rooflight[1]	Maximum total area of diffuser panels and rooflights as percentage of floor area of the space in which ceiling is located	Minimum separation distance between diffuser panels or rooflights[1]
TP(a)	Any except protected stairway	No limit[2]	No limit	No limit
Class 3[3] or TP(b)	Rooms	5 m²	50%[4]	3 m
	Circulation spaces except protected stairways	5 m²	15%[4]	3 m

Notes:
1 Smaller panels can be grouped together provided that the overall size of the group and the space between one group and any others satisfies the dimensions shown in Figure C.2.
2 Lighting diffusers of TP(a) flexible rating should be restricted to panels of not more than 5 m² each, see paragraph 3.14.
3 There are no limitations on Class 3 material in small rooms.
4 The minimum 3 m separation specified in Figure C.2 between each 5 m² must be maintained. Therefore, in some cases it may not be possible to use the maximum percentage quoted.

Approved Document B3: Internal fire spread (structure)

D

For Scotland see Appendix E of this document.

For Northern Ireland see the relevant deemed-to-satisfy provisions of the publications of Technical Booklet E:2005 as referred to in Table E to Schedule 5 of the Building Regulations (Northern Ireland) 2000 as amended. The legislation and technical booklet can be found on the Department of Finance and Personnel's building regulations website: www.buildingregulationsni.gov.uk

Note: Approved Document B (Fire safety) has been subdivided into two volumes. Volume 1 contains guidance applicable to dwellinghouses while Volume 2 contains guidance applicable to buildings other than dwellinghouses. As such, the guidance below is similarly divided along these lines.

In the text below, Approved Document B is abbreviated to ADB.

Volume 1: Dwellinghouses

Note: Volume 1 is relevant to houses and the individual residential units of sheltered housing. For guidance applicable to the common parts of a sheltered housing development, flats, student accommodation and other buildings, reference should be made to Volume 2.

Volume 2: Buildings other than dwellinghouses

Note: Volume 2 is relevant to buildings other than dwellinghouses. For guidance applicable to dwellinghouses and the individual dwellings forming part of controlled sheltered housing, reference should be made to Volume 1.

7.2 (Vol. 1) and 10.2 (Vol. 2)
If a fire-separating element of construction such as a ceiling, floor, wall or similar is to be effective, it is essential that every joint, imperfection of fit or, of particular interest to this publication, opening to allow services to pass through such elements should be adequately sealed or fire-stopped such that the fire resistance of the element is not reduced.

7.3 (Vol. 1) and 10.3 (Vol. 2)

The measures described in Section 7 of ADB Vol. 1 and Section 10 of ADB Vol. 2 'Protection of openings and fire-stopping' are primarily intended to delay the passage of fire through a building. It is accepted that these measures will additionally retard smoke spread through a building. However, there are no specific criteria relating to passage of smoke.

7.4 (Vol. 1)

Consideration should be given to the effect that the installation of electrical equipment and accessories including downlighters may have in reducing the fire resistance of elements of the construction. In some cases some additional protective measures may be required.

7.12 (Vol. 1) and 10.17 (Vol. 2)

Fire-stopping is required in all

▶ joints between fire-separating elements of the construction
▶ openings made to allow pipes, ducts, conduits, trunking or cables to pass through fire-separating elements of the construction.

Any such openings for building services should be as physically small as possible and their numbers kept to a minimum.

7.13 (Vol. 1) and 10.18 (Vol. 2)

In order to prevent displacement and hence reduced effectiveness, any materials used for fire-stopping should be reinforced with, or be supported by, materials having limited combustibility:

▶ where the unsupported span is greater than 100 mm; and
▶ in any other case where non-rigid materials are used (unless it has been demonstrated by test that they are suitable without such reinforcement).

7.14 (Vol. 1) and 10.19 (Vol. 2)

A number of proprietary fire-stopping and sealing products, some of which are specifically designed for service penetrations, that have been shown by test to maintain fire resistance of construction elements are available. These may be used in accordance with manufacturers' instructions.

Other materials which may be employed for fire-stopping and sealing include:

▶ cement-based or gypsum-based vermiculite/perlite mixes
▶ cement mortar
▶ glass fibre-, crushed rock-, blast furnace slag-, or ceramic-based products (with or without additional binding elements)
▶ gypsum-based plaster
▶ intumescent mastics.

It should be recognised that not all of the above will be suitable for all situations and an appropriate material should be selected to suit particular individual circumstances.

Reference should also be made to section 6.2 of this publication.

Scottish Technical Standards relating to protection against fire

E

For England and Wales see Appendices B, C and D of this document.

For Northern Ireland see the relevant deemed-to-satisfy provisions of the publications of Technical Booklet E:2005 as referred to in Table E to Schedule 5 of the Building Regulations (Northern Ireland) 2000 as amended. The legislation and technical booklet can be found on the Department of Finance and Personnel's building regulations website: www.buildingregulationsni.gov.uk

The Scottish Building Standards Division (BSD) have produced two Technical Handbooks intended to provide guidance on achieving the standards set in the Building (Scotland) Regulations 2004. One Technical Handbook relates to Domestic buildings and the other to Non-domestic buildings.

Compliance with the Building (Scotland) Regulations 2004 can be achieved by compliance with the mandatory Building Standards and the associated guidance given in the Technical Handbooks.

The last revision of the Technical Handbooks came into force on 4 January 2009.

The information contained in this appendix relates to Section 2 'Fire' and Section 4 'Safety' of the Technical Handbooks.

The number of the Mandatory Standard being discussed will appear in the margin along with an indication of whether the Mandatory Standard is applicable to domestic and/or non-domestic buildings.

2.1 Compartmentation

Every building must be designed and constructed in such a way that in the event of an outbreak of fire within the building, fire and smoke are inhibited from spreading beyond the compartment of origin until any occupants have had the time to leave that compartment and any fire containment measures have been initiated.

Limitation:
This standard does not apply to domestic buildings.

2.1.14 Non-domestic

This section deals with openings and service penetrations.

Any openings or service penetrations in elements of the building structure should be kept to a minimum.

Self-closing fire doors can be fitted with hold open devices as specified in BS 5839-3: 1998 *Fire detection and fire alarm systems for buildings. Specification for automatic release mechanisms for certain fire protection equipment* provided that the door is not an emergency door, a protected door serving the only escape stair in the building (or the only escape stair serving part of the building) or a protected door serving a fire-fighting shaft.

If it is decided to fit automatic release mechanisms as described above, this may make it necessary to install an automatic fire detection/alarm system.

Any service opening other than a ventilating duct which penetrates a compartment wall or compartment floor should be fire-stopped such that it provides at least the same duration of fire resistance as if the wall or floor had not been so penetrated. This may be achieved by:

▶ a casing which has at least the appropriate fire resistance from the outside; or
▶ a casing which has at least half the appropriate fire resistance from each side; or
▶ an automatic heat-activated sealing device that will maintain the appropriate fire resistance in respect of integrity for the wall or floor regardless of the opening size.

Fire-stopping need not be provided for:

▶ a pipe or a cable with a bore, or diameter, of not more than 40 mm; or
▶ not more than four 40 mm diameter pipes or cables that are at least 40 mm apart and at least 100 mm from any other pipe; or
▶ more than four 40 mm diameter pipes or cables that are at least 100 mm apart.

Fire-stopping may be required to close an imperfection of fit or design tolerance between construction elements and components, around service openings and ventilation ducts. Any proprietary fire-stopping products, including intumescent products, should be subjected to testing to demonstrate their ability to maintain a sufficient degree of fire resistance under the conditions in which they are to be employed.

Where only minimal differential movement is anticipated, either in normal use or during fire exposure, the following may be used:

- proprietary fire-stopping products;
- cement mortar;
- gypsum-based plaster;
- cement- or gypsum-based vermiculite/perlite mixes;
- mineral fibre, crushed rock and blast furnace slag or ceramic-based products (with or without added binders).

Where greater differential movement is anticipated, either in normal use or during fire exposure, proprietary fire-stopping products may be used.

In order to prevent their displacement, materials used for fire-stopping should be provided with reinforcement or support from non-combustible materials where the unsupported span is more than 100 mm and where non-rigid materials are used. This will not be necessary where it has been shown by test that the materials alone are satisfactory within their field of application.

Reference should also be made to section 6.2 of this publication.

2.2 Separation

> Every building that is divided into more than one area of different occupation must be designed and constructed in such a way that in the event of an outbreak of fire within the building, fire and smoke are inhibited from spreading beyond the area of occupation where the fire originated.

Domestic and non-domestic

2.2.4 Non-domestic and 2.2.7 Domestic
In order to reduce the risk of a fire starting within a combustible separating wall or a fire spreading rapidly on or within the wall construction:

- insulation material exposed in a cavity should be of low risk or non-combustible materials (see annex 2.E); and
- the internal wall lining should be constructed from material which is low risk or non-combustible; and
- the wall should contain no pipes, wires or other services.

2.2.9 Domestic
Any openings or service penetrations in elements of the building structure should be kept to a minimum.

Self-closing fire doors can be fitted with hold open devices as specified in BS 5839-3: 1998 *Fire detection and fire alarm systems for buildings. Specification for automatic release mechanisms for certain fire protection equipment* provided that the door is not an emergency door, a protected door serving the only escape stair in the building (or the only escape stair serving part of the building) or a protected door serving a fire-fighting shaft.

If it is decided to fit automatic release mechanisms as described above, this may make it necessary to install an automatic fire detection/alarm system.

Any service opening other than a ventilating duct which penetrates a separating wall or separating floor should be fire-stopped such that it provides at least the same duration of fire resistance as if the wall or floor had not been so penetrated. This may be achieved by:

▶ a casing which has at least the appropriate fire resistance from the outside; or
▶ a casing which has at least half the appropriate fire resistance from each side; or
▶ an automatic heat-activated sealing device that will maintain the appropriate fire resistance in respect of integrity for the wall or floor regardless of the opening size.

Fire stopping need not be provided for:

▶ a pipe or a cable with a bore, or diameter, of not more than 40 mm; or
▶ not more than four 40 mm diameter pipes or cables that are at least 40 mm apart and at least 100 mm from any other pipe; or
▶ more than four 40 mm diameter pipes or cables that are at least 100 mm apart.

Fire-stopping may be required to close an imperfection of fit or design tolerance between construction elements and components, around service openings and ventilation ducts. Any proprietary fire-stopping products, including intumescent products, should be subjected to testing to demonstrate their ability to maintain a sufficient degree of fire resistance under the conditions in which they are to be employed.

Where only minimal differential movement is anticipated, either in normal use or during fire exposure the following may be used:

▶ proprietary fire-stopping products;
▶ cement mortar;
▶ gypsum-based plaster;
▶ cement- or gypsum-based vermiculite/perlite mixes;
▶ mineral fibre, crushed rock and blast furnace slag or ceramic-based products (with or without added binders).

Where greater differential movement is anticipated, either in normal use or during fire exposure, proprietary fire-stopping products may be used.

In order to prevent their displacement, materials used for fire-stopping should be provided with reinforcement or support from non-combustible materials where the unsupported span is more than 100 mm and where non-rigid materials are used. This will not be necessary where it has been shown by test that the materials alone are satisfactory within their field of application.

Reference should also be made to section 6.2 of this publication.

2.3 Structural protection

Domestic and non-domestic

> Every building must be designed and constructed in such a way that in the event of an outbreak of fire within the building, the load-bearing capacity of the building will continue to function until all occupants have escaped, or been assisted to escape, from the building and any fire containment measures have been initiated.

2.3.4 Domestic and non-domestic

In general, openings and service penetrations in elements of structure do not need to be protected from fire unless there is the possibility of structural failure.

Where a large opening or a large number of small openings are made, care should be taken to ensure that the load-bearing capacity of the structure is maintained.

2.4 Cavities

Every building must be designed and constructed in such a way that in the event of an outbreak of fire within the building, the unseen spread of fire and smoke within concealed spaces in its structure and fabric is inhibited.

Domestic and non-domestic

2.5 Internal linings

Every building must be designed and constructed in such a way that in the event of an outbreak of fire within the building, the development of fire and smoke from the surfaces of walls and ceilings within the area of origin is inhibited.

Domestic and non-domestic

2.5.7 Domestic and non-domestic

Thermoplastic materials may be used in light fittings with diffusers. Where the lighting diffuser forms an integral part of the ceiling, the size and disposition of the lighting diffusers should be installed in accordance with the recommendations given in Table E.1 and Figure E.1.

▼ Table E.1
Thermoplastic rooflights and light fittings with diffusers

Classification of lower surface	Protected zone or fire-fighting shaft	Unprotected zone or protected enclosure		Room	
	Any thermoplastic	TP(a) rigid	TP(a) flexible and TP(b)	TP(a) rigid	TP(a) flexible and TP(b)
Maximum area of each diffuser panel or rooflight	Not advised	No limit	5 m²	No limit	5 m²
Maximum total area of diffuser panels or rooflights as a percentage of the floor area of the space in which the ceiling is located	Not advised	No limit	15%	No limit	50%
Minimum separation distance between diffuser panels or rooflights	Not advised	No limit	3 m	No limit	3 m

Notes:

1 Smaller panels can be grouped together provided that the overall size of the group, and the space between any others, satisfies the dimensions shown in Figure E.1.

2 The minimum 3 m separation in Figure E.1 should be maintained between each 5 m² panel. In some cases, therefore, it may not be possible to use the maximum percentage quoted.

3 TP(a) flexible is not recommended in rooflights.

Where the lighting diffusers form an integral part of a fire-resisting ceiling which has been satisfactorily tested, the amount of thermoplastic material is unlimited.

Where light fittings with thermoplastic diffusers do not form an integral part of the ceiling, the amount of thermoplastic material is unlimited provided the lighting diffuser is designed to fall out of its mounting when softened by heat.

▼ **Figure E.1**
Layout restrictions on thermoplastic rooflights and light fittings with diffusers

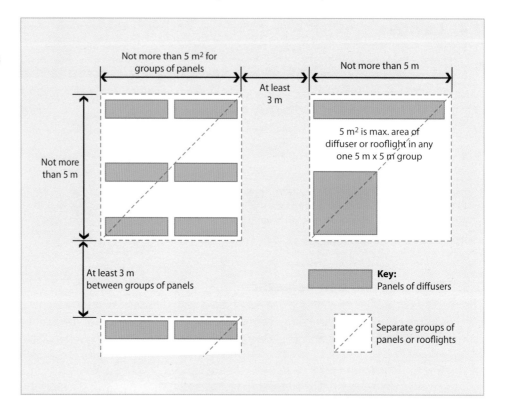

2.9 Escape

Domestic and non-domestic

Every building must be designed and constructed in such a way that in the event of an outbreak of fire within the building, the occupants, once alerted to the outbreak of the fire, are provided with the opportunity to escape from the building, before being affected by fire or smoke.

2.10 Escape lighting

Domestic and non-domestic

Every building must be designed and constructed in such a way that in the event of an outbreak of fire within the building, illumination is provided to assist in escape.

2.10.0 Domestic
Escape lighting is not required for dwellings as it is assumed that occupants will be familiar with the layout. In buildings containing flats and maisonettes, any common escape routes will require illumination to assist occupants to find their way to a place of safety.

2.10.0 Non-domestic

Escape routes in non-domestic premises should be illuminated to aid the safe evacuation of a building in an emergency.

It may be necessary to provide emergency lighting and exit signage under the Fire Safety (Scotland) Regulations 2006 (for further information, refer to clause 2.0.4 of the non-domestic Technical Handbook).

Part 1 of the Cinematograph (Safety) (Scotland) Regulations 1955 contains specific requirements for lighting for buildings such as cinemas.

Additional guidance on special fire precautions that may be required within residential care buildings, hospitals and enclosed shopping centres is grouped in the annexes to the non-domestic Technical Handbook as follows:

▶ residential care buildings, see annex 2.A
▶ hospitals, see annex 2.B
▶ enclosed shopping centres, see annex 2.C

These annexes should help designers and verifiers to find the information they require quickly when designing or vetting such buildings.

It should be remembered that the guidance in the annexes is in addition and supplementary to the guidance to standard 2.1 to 2.15.

2.10.1 Domestic and non-domestic

Escape routes should be provided with artificial lighting supplied by a protected circuit to provide a level of illumination not less than that recommended for emergency lighting (see clause 2.10.3). Where artificial lighting serves a protected zone, it should be supplied from a protected circuit separate from that supplying any other part of the escape route.

Artificial lighting supplied by a protected circuit need not be provided where a system of emergency lighting is installed.

2.10.2 Domestic and non-domestic

A protected circuit is a circuit originating at the main incoming switch or distribution board, the conductors of which are protected against fire.

Regardless of what system is employed, escape routes should be capable of being illuminated when the building is in use.

2.10.3 Domestic and non-domestic

Emergency lighting is lighting designed to come into, or remain in, operation automatically in the event of a local and general power failure and should be installed in buildings considered to be at higher risk, such as in high-rise buildings, buildings with basements or in rooms where the number of people is likely to exceed 60.

Emergency lighting should be installed in buildings or parts of a building considered to be at higher risk such as:

▶ in a protected zone and an unprotected zone in a building with any storey at a height of more than 18 m;
▶ in a room with an occupancy capacity of more than 60, or in the case of an inner room where the combined occupancy capacity of the inner room plus

the adjoining room (and any protected zone or unprotected zone serving these rooms) is more than 60;

▶ in an underground car park including any protected zone or unprotected zone serving it where less than 30 per cent of the perimeter of the car park is open to the external air;

▶ in a protected zone or unprotected zone serving a basement storey;

▶ in a place of special fire risk (other than one requiring access only for the purposes of maintenance) and any protected zone or unprotected zone serving it;

▶ in any part of an air supported structure;

▶ in a protected zone or unprotected zone serving a storey which has at least two storey exits in the following buildings:

– entertainment, assembly, factory, shop, multi-storey storage (Class 1), single-storey storage (Class 1) with a floor area more than 500 m²;

– a protected zone or unprotected zone serving a storey in a multi-storey non-residential school;

– a protected zone or unprotected zone serving any storey in an open-sided car park.

Emergency lighting in places of entertainment, such as cinemas, bingo halls, ballrooms, dance halls and bowling alleys, should be in accordance with CP 1007:1955. Emergency lighting in any other building should be in accordance with BS 5266: Part 1:2005 as read in association with BS 5266: Part 7:1999 (BS EN 1838:1999).

In the case of a building with a smoke and heat exhaust ventilation system, the emergency lighting should be sited below the smoke curtains or installed so that it is not rendered ineffective by smoke-filled reservoirs.

2.11 Communication

Domestic and non-domestic

> Every building must be designed and constructed in such a way that in the event of an outbreak of fire within the building, the occupants are alerted to the outbreak of fire.
>
> **Limitation:**
> This standard applies only to a building which is:
> **a** a dwelling;
> **b** a residential building; or
> **c** an enclosed shopping centre.

2.11.0 Domestic

It is accepted that early detection and warning of fire to occupants of domestic premises can play a vital role in increasing their chances of escape. This is particularly important given that the occupants will be asleep some of the time.

2.11.0 Non-domestic

It is recommended that automatic fire detection is installed in residential buildings and enclosed shopping centres, as the risk to life is far greater in residential buildings because the occupants may be asleep or, in the case of enclosed shopping centres, large numbers of the public may be present who could be unfamiliar with the building and may need to travel long distances in order to leave the building safely.

Automatic fire detection systems need not be installed in all other buildings other than:

▶ to compensate for some departure from the guidance elsewhere in this handbook;

▶ as part of the operating system for some fire protection systems, such as pressure differential systems (see clause 2.9.16), automatic door release devices (see clause 2.1.14) or electrically operated locks (see clause 2.9.15);

▶ a building designed on the basis of phased evacuation (see clause 2.9.11).

2.11.1 Domestic

Dwellings having no storey exceeding 200 m² should be provided with one or more smoke alarms located on each storey. The smoke alarms should be installed in accordance with the guidance given in clause 2.11.2 of the Technical Handbook.

2.11.2 Domestic

The standby power supply for the smoke alarm should be from a primary battery, a secondary battery or a capacitor. The capacity of the standby supply should be sufficient to power the smoke alarm for at least 72 hours when the mains power supply is off, while giving a visual warning of the mains power supply being off. Sufficient capacity should remain to provide a warning of smoke for a further 4 minutes.

Where the capacity of the standby power supply falls below the recommended standby duration when the mains power supply is on, an audible warning should be given at least once every minute. This warning cycle should persist for at least 30 days when the mains power supply is off.

Smoke alarms should be ceiling mounted and located:

▶ in circulation areas which will be used as a route along which to escape, not more than 7 m from the door to a living room or kitchen and not more than 3 m from the door to a room intended to be used as sleeping accommodation, the dimensions to be measured horizontally;

▶ where the circulation area is more than 15 m long, not more than 7.5 m from another smoke alarm on the same storey;

▶ at least 300 mm away from any wall or light fitting, heater or air conditioning outlet;

▶ on a surface which is normally at the ambient temperature of the rest of the room or circulation area in which the smoke alarm is situated.

The above recommendations are broadly in line with the recommendations of BS 5839: Part 6:2004 for a Grade D Type LD3 system.

Where more than one smoke alarm is installed in a dwelling they should be interconnected so that detection of a fire by any one of them operates the alarm signal in all of them.

A smoke alarm should be permanently wired to a circuit. The mains supply to the smoke alarm should take the form of either:

▶ an independent circuit at the dwelling's main distribution board, in which case no other electrical equipment should be connected to this circuit (other than a dedicated monitoring device installed to indicate failure of the mains supply to the smoke alarms); or

▶ a separately electrically protected, regularly used local lighting circuit.

Smoke alarms may be interconnected by 'hard wiring' on a single final circuit.

Any smoke alarm in a dwelling which forms part of residential accommodation with a warden or supervisor, should have a connection to a central monitoring unit so that in the event of fire the warden or supervisor can identify the dwelling concerned, and the system should follow the guidance in BS 5839: Part 6:2004 for a Grade C Type LD2 installation.

In order to reduce the frequency of unwanted false alarms, guidance is provided in BS 5839: Part 6:2004 on the types of sensor most appropriate for the circumstances.

2.11.3 Domestic
Dwellings having any storey area more than 200 m^2 should be provided with a fire detection and alarm system designed and installed in accordance with BS 5839 Part 6:2004 for a Grade C Type LD2 installation.

2.11.1 Non-domestic
Residential buildings contain occupants who may be asleep. This presents particular problems in the event of an outbreak of fire as the occupants will not be aware that their lives may be at risk. A higher level of automatic fire detection coverage is recommended in residential care buildings and hospitals to give occupants and staff the earliest possible warning of an outbreak of fire and allow time for assisting occupants to evacuate the building in an emergency (see Annex 2.A and Annex 2.B).

It is accepted that early warning of fire in residential buildings increases significantly the level of safety of the occupants. For this reason, residential buildings (other than residential care buildings and hospitals) should be provided with an automatic fire detection and alarm system installed in accordance with the following recommendations:

▶ automatic detection is installed to at least Category L2 in BS 5839: Part 1:2002;
▶ manual fire alarm call points as specified in BS EN 54-11: 2001 (Type A) should be sited in accordance with BS 5839: Part 1:2002;
▶ the fire alarm is activated upon the operation of manual call points, automatic detection or the operation of any automatic fire suppression system installed;
▶ the audibility level of the fire alarm sounders should be as specified in BS 5839: Part 1:2002.

In the case of shared residential accommodation designed to provide sleeping accommodation for not more than six persons and having no sleeping accommodation below ground level or above first floor level, a domestic system comprising smoke alarms may be installed in accordance with the domestic Technical Handbook.

2.11.2 Non-domestic
Guidance relating to fire safety in residential care buildings is given in Annex 2.A of the non-domestic Technical Handbook.

2.11.3 Non-domestic
Guidance relating to fire safety in hospitals is given in Annex 2.B of the non-domestic Technical Handbook.

2.11.4 Non-domestic

The distance people need to travel to reach a protected zone or a place of safety in enclosed shopping centres will be extended due to their having a unique design. Alternative life safety systems should be installed as a means of providing compensation for the extended travel distances and the lack of compartmentation within the building. Automatic fire detection systems in enclosed shopping centres form an integral part of the overall life safety strategy in such buildings and should be compatible with, and interact with, other mechanical and electrical equipment.

Guidance on the selection of fire detection/alarm systems in enclosed shopping centres is given in Annex 2.C of the non-domestic Technical Handbook.

4.5   Electrical safety

Every building must be designed and constructed in such a way that the electrical installation does not:

Domestic and
non-domestic

a   threaten the health and safety of the people in and around the building; and
b   become a source of fire.

Limitation:
This standard does not apply to an electrical installation:

a   serving a building or any part of a building to which the Mines and Quarries Act 1954 or the Factories Act 1961 applies; or
b   forming part of the works of an undertaker to which regulations for the supply and distribution of electricity are made under the Electricity Act 1989.

4.5.0 Domestic and non-domestic

The hazards posed by unsafe electrical installation are injuries caused by contact with electricity (shocks and burns) and injuries arising from fires in buildings ignited through malfunctioning or incorrect installations.

The intention of this standard is to ensure that electrical installations are safe in terms of the hazards likely to arise from defective installations, namely fire, electric shock and burns or other personal injury.

Installations should:

▶   safely accommodate any likely maximum demand; and
▶   incorporate appropriate automatic devices for protection against overcurrent or leakage; and
▶   provide means of isolating parts of the installation or equipment connected to it, as are necessary for safe working and maintenance.

The standard applies to fixed installations in buildings. An installation consists of the electrical wiring and associated components and fittings, including all permanently secured equipment, but excluding portable equipment and appliances.

Index

S

T

W

The

IEE Wiring Regulations and associated publications

The IEE prepares regulations for the safety of electrical installations for buildings, the *IEE Wiring Regulations* (BS 7671 *Requirements for Electrical Installations*), which have now become the standard for the UK and many other countries. It also recommends, internationally, the requirements for ships and offshore installations. The IEE provides guidance on the application of the installation regulations through publications focused on the various activities, from design of the installation through to final test and then maintenance. This includes a series of eight Guidance Notes, two Codes of Practice and Model Forms for use in Wiring Installations.

**Requirements for Electrical Installations
BS 7671:2008 (IEE Wiring Regulations,
17th Edition)**
Order book PWR1700B Paperback 2008
ISBN: 978-0-86341-844-0   **£65**

On-Site Guide (BS 7671:2008 17th Edition)
Order book PWGO170B 188pp Paperback 2008
ISBN: 978-0-86341-854-9   **£20**

Wiring Matters Magazine   **FREE**
If you wish to receive a FREE copy or advertise
in Wiring Matters please visit
www.theiet.org/wm

IEE Guidance Notes

A series of Guidance Notes has been issued,
each of which enlarges upon and amplifies
the particular requirements of a part of the IEE
Wiring Regulations.

**Guidance Note 1: Selection & Erection of
Equipment, 5th Edition**
Order book PWG1170B 216pp Paperback 2009
ISBN: 978-0-86341-855-6   **£30**

**Guidance Note 2: Isolation & Switching,
5th Edition**
Order book PWG2170B 74pp Paperback 2009
ISBN: 978-0-86341-856-3   **£25**

**Guidance Note 3: Inspection & Testing,
5th Edition**
Order book PWG3170B 128pp Paperback 2008
ISBN: 978-0-86341-857-0   **£25**

**Guidance Note 4: Protection Against Fire,
5th Edition**
Order book PWG4170B 98pp Paperback 2009
ISBN: 978-0-86341-858-7   **£25**

**Guidance Note 5: Protection Against
Electric Shock, 5th Edition**
Order book PWG5170B 115pp Paperback 2009
ISBN: 978-0-86341-859-4   **£25**

**Guidance Note 6: Protection Against
Overcurrent, 5th Edition**
Order book PWG6170B 113pp Paperback 2009
ISBN: 978-0-86341-860-0   **£25**

**Guidance Note 7: Special Locations,
3rd Edition**
Order book PWG7170B 142pp Paperback 2009
ISBN: 978-0-86341-861-7   **£25**

**Guidance Note 8: Earthing & Bonding,
1st Edition**
Order book PWRG0241 168pp Paperback 2007
ISBN: 978-0-86341-616-3   **£25**

continues overleaf ▶

Other guidance publications

**Commentary on IEE Wiring Regulations
(17th Edition, BS 7671:2008)**
Order book PWR08640
c.400pp Hardback 2009
ISBN: 978-0-86341-966-9 **£65**

Electrical Maintenance, 2nd Edition
Order book PWR05100
227pp Paperback 2006
ISBN: 978-0-86341-563-0 **£35**

**Code of Practice for In-service Inspection and
Testing of Electrical Equipment, 3rd Edition**
Order book PWR08630
138pp Paperback 2007
ISBN: 978-0-86341-833-4 **£35**

**Electrical Craft Principles, Volume 1,
5th Edition**
Order book PBNS0330
344pp Paperback 2009
ISBN: 978-0-86341-932-4 **£25**

**Electrical Craft Principles, Volume 2,
5th Edition**
Order book PBNS0340
432pp Paperback 2009
ISBN: 978-0-86341-933-1 **£25**

**Electrician's Guide to the Building
Regulations, 2nd Edition**
Order book PWGP170B
234pp Paperback 2008
ISBN: 978-0-86341-862-4 **£20**

**Electrical Installation Design Guide:
Calculations for Electricians and Designers**
Order book PWR05030
186pp Paperback 2008
ISBN: 978-0-86341-550-0 **£20**

Electrician's Guide to Emergency Lighting
Order book PWR05020
88pp Paperback 2009
ISBN: 978-0-86341-551-7 **£20**

Electrical training courses

We offer a comprehensive range of technical training at
many levels, serving your training and career development
requirements as and when they arise.

Courses range from Electrical Basics to Qualifying City &
Guilds or EAL awards.

Train to the 17th Edition BS 7671:2008

▶ Update from 16th to 17th Edition
▶ Understand the changes
▶ New qualifying awards C&G/EAL
▶ Meet industry standards

Qualifying Courses

▶ Certificate of Competence Management of Electrical
  Equipment Maintenance (PAT) – 1 day
▶ Certificate of Competence for the Inspection and
  Testing of Electrical Equipment (PAT) – 1 day
▶ Certificate in the Requirements for Electrical
  Installations – 3 days
▶ Upgrade from 16th Edition achieved since 2001 –
  1 day
▶ Certificate in Fundamental Inspection, Testing and
  Internal Verification – 3 days
▶ Certificate in Inspection, Testing and Certification of
  Electrical Installations – 3 days

Other 17th Edition Courses

▶ Earthing & Bonding – For designers and electrical
  contractors who require a good working knowledge
  of the E & B arrangements as required by
  BS 7671:2008
▶ 17th Edition Design – BS 7671 and the principles
  associated with the design of electrical installations

To view all our current courses and book online, visit
www.theiet.org/coursesbr

**To discuss your training requirements and for on-
site group training, please speak to one of our
advisors on +44 (0)1438 767289**

Collective **inspiration**

Order Form

Details

Name:

Job Title:

Company/Institution:

Address:

Postcode: Country:

Tel: Fax:

Email:

Membership No (if Institution member):

Payment methods

☐ By **cheque** made payable to The Institution of Engineering and Technology

☐ By **credit/debit card:**

☐ Visa ☐ Mastercard ☐ American Express ☐ Maestro Issue No:_____

Valid from: ☐☐ ☐☐ Expiry Date: ☐☐ ☐☐ Card Security Code: ☐☐☐☐
(3 or 4 digits on reverse of card)

Card No: ☐☐☐☐ ☐☐☐☐ ☐☐☐☐ ☐☐☐☐

Signature_____ Date _____
(Orders not valid unless signed)

Cardholder Name:

Cardholder Address:

Town: Postcode:

Country:

☐ By official **company purchase order** (please attach copy)
EU VAT number:_____

Ordering information

Quantity	Book No.	Title/Author	Price (£)
		Subtotal	
		- Member discount**	
		+ Postage /Handling*	
		+ VAT (if applicable)	
		Total	

Membership

Passionate about engineering? Committed to your career?

Do you want to join an organisation that is inspiring, insightful and innovative?

One of the most highly recognised knowledge sharing networks in the world, membership to the Institution of Engineering and Technology is for engineers and technologists working or studying in an increasingly multidisciplinary, digital and global environment.

Joining the IET and having access to tailored products and services will become invaluable for your career and can be your first step towards professional qualifications.

You could take advantage of ...

- ▶ Fortnightly copy of the industry's leading publication, *Engineering & Technology* magazine.

- ▶ Professional development and career support services to help gain registration.

- ▶ Dedicated training courses, seminars and events covering a wide range of subjects and skills.

- ▶ Watch live IET.tv event footage at your desktop via the internet, ask the speaker questions during live streaming and feel part of the audience without physically being there.

- ▶ Access to over 100 local networks around the world.

- ▶ Meet like-minded professionals through our array of specialist online communities.

- ▶ Instant online access to over 70,000 books, 3,000 periodicals and full-text collections of electronic articles – wherever you are in the world.

- ▶ Discounted rates on IET books and technical proceedings.

Join online today www.theiet.org/join or contact our membership and customer service centre on +44 (0)1438 765678

Professional Registration

What type of registration is for you?

Chartered Engineers (CEng) develop appropriate solutions to engineering problems, using new or existing technologies, through innovation, creativity and change. They might develop and apply new technologies, promote advanced designs and design methods, introduce new and more efficient production techniques, marketing and construction concepts, pioneer new engineering services and management methods. Chartered Engineers are engaged in technical and commercial leadership and possess interpersonal skills.

Incorporated Engineers (IEng) maintain and manage applications of current and developing technology, and may undertake engineering design, development, manufacture, construction and operation. Incorporated Engineers are engaged in technical and commercial management and possess effective interpersonal skills.

Engineering Technicians (EngTech) are involved in applying proven techniques and procedures to the solution of practical engineering problems. You will carry supervisory or technical responsibility, and are competent to exercise creative aptitudes and skills within defined fields of technology. Engineering Technicians also contribute to the design, development, manufacture, commissioning, operation or maintenance of products, equipment, processes or services.

For further information on Professional Registration (CEng/IEng/EngTech), tel: +44 (0)1438 767282 or email: membership@theiet.org

Notes

Notes

Notes

Notes